Boolean Gröbner bases – Theory, Algorithms and Applications

Boolesche Gröbnerbasen –
Theorie, Algorithmen, Anwendungen

Vom Fachbereich Mathematik
der Technischen Universität Kaiserslautern
zur Verleihung des akademischen Grades
Doktor der Naturwissenschaften
(Doctor rerum naturalium, Dr.)
genehmigte

Dissertation

von

Dipl.-Math. Michael Brickenstein

aus Hamburg

Kaiserslautern 2009

D 386

1. Gutachter: Prof. Dr. Dr. h. c. Gert-Martin Greuel,
Technische Universität Kaiserslautern
2. Gutachter: Univ.-Prof. Dr. Armin Biere,
Johannes Kepler Universität Linz
Datum der Disputation: 19. Februar 2010

Bibliografische Information der Deutschen Nationalbibliothek

Die Deutsche Nationalbibliothek verzeichnet diese Publikation in der Deutschen Nationalbibliografie; detaillierte bibliografische Daten sind im Internet über http://dnb.d-nb.de abrufbar.

©Copyright Logos Verlag Berlin GmbH 2010
Alle Rechte vorbehalten.

ISBN 978-3-8325-2597-2

Logos Verlag Berlin GmbH
Comeniushof, Gubener Str. 47,
10243 Berlin
Tel.: +49 (0)30 42 85 10 90
Fax: +49 (0)30 42 85 10 92
INTERNET: http://www.logos-verlag.de

Contents

Introduction

There exist very few concepts in computational algebra which are as central for theory and have as many applications as Gröbner bases. Since Bruno Buchberger had introduced the concept in 1965, the community did much research on the topic. Algorithmic enhancements have been the driving force that facilitated many results in other areas.

The work presented in this thesis includes a small part about general Gröbner bases computations in commutative and noncommutative polynomial rings. This section presents the "slimgb" algorithm, originally introduced in the author's diploma thesis [Brickenstein, 2004], in a generalized, enhanced and more mature form. Meanwhile, the description of the algorithm is accepted for publication [Brickenstein, 2010]. Since "slimgb" is publicly available in the computational algebra system SINGULAR, a variety of researchers have used it. Thereby, they were able to tackle larger problems that state of the art algorithms and implementations could not solve in reasonable time before. Here, the following publications should be mentioned: King [2007], Levandovskyy and Morales [2008].

The main part concerns more recent research specialized on Boolean polynomial rings: It treats the special – but nevertheless important – case of polynomial equation systems over $\mathbb{Z}_2$. Furthermore, we restrict the solutions to $\mathbb{Z}_2$ itself. For this purpose, we append field equations of the form $x^2 = x$ to the equation system. Thus, the complete system of equations can be expressed by the set of field polynomials and the so called Boolean polynomials: polynomials over $\mathbb{Z}_2$ where each term has a degree bound of one per variable. There exist many direct applications for example from formal verification and cryptanalysis for this setting. Other problems like polynomial equations of $\mathbb{Z}_2^r$ with solutions in $\mathbb{Z}_2^r$ can be reformulated as problems with Boolean polynomials.

Formal verification problems like the one given in [Krautz et al., 2008] inspired our research. Formal verification is a domain where much research was done. In many settings, techniques from satisfiability theory [Biere et al., 2009] can be applied. SAT-solvers are specialized tools for problems in propositional logic. Nowadays, there exist high-performance implementations using excellent heuristics.

This thesis and the POLYBORI framework apply Gröbner basis theory as a different approach. The latter was developed in a much more general theoretical setting than Boolean computations. One can hope that at some point in time computational algebra might form a reasonable supplement of the existing and well developed techniques in formal verification. Since the mapping from propositional logic to Boolean polynomials is quite natural, there have been several approaches already in the past for describing integrated circuits in terms of commutative algebra [Agnarsson et al., 1984]. Gröbner bases form a standard tool for computations in the algebraic domain. Still, the various methods using propositional logic cannot efficiently handle all problems. Hence, several research groups made considerable efforts to use Gröbner bases for formal verification and satisfiability. These past approaches focussed on various points: Tran and Vardi [2007] applied ideal theory to Symbolic Model Checking. Clegg, Edmonds, and Impagliazzo [1996] combined the Buchberger algorithm with backtracking. Condrat and Kalla [2007] used Gröbner bases as preprocessing step for small subsystems. Our approach is quite orthogonal to the previous ones. We combine techniques originating from formal verification itself (decision diagrams) with algorithmic research on Gröbner bases.

The PolyBoRi developers (including myself) have always tried to be very open to as many applications as possible. Therefore, we have published the software quite early and supported the integration into the Sage mathematics software [Stein et al., 2009]. By this, many researchers are able to use it.

Since Boolean polynomials have many applications, they have already been subject to scientific research. Tran and Vardi [2007] claim to have proven that the Gröbner bases computations for Boolean polynomial rings have complexity of polynomial space consumption. However polynomial space consumption of the algorithm does not imply polynomial running time. Another important result given in [Bardet et al., 2005] was that semi-regular sequences can have better complexities than the general case. It forms the theoretical foundation of the algebraic HFE attacks [Faugère, 2003].

The concepts of a Boolean polynomial and Boolean Gröbner bases appear already in [Sakai and Sato, 1988]. Our notion of Boolean polynomials is equivalent to the one which Sakai and Sato describe as Boolean normal polynomial. Sato, Nagai, and Inoue [2008a] modify and generalize this concept to quotient rings of the polynomial ring with field polynomial $x^2 = x$ where the coefficient ring is not necessarily a field. It may be a more general commutative coefficient ring that also satisfies $c^2 + c = 0$ for each c. The searched literature applies only one coefficient ring that differs from $\mathbb{Z}_2$ and was used with that concept of Boolean rings: $\mathbb{Z}_2^r$ [Sato, Suzuki, Inoue, and Nabeshima, 2008b]. These problems can also be expressed as problems for our more basic Boolean polynomials, when considering modules. Of course Gröbner bases over modules can be computationally modeled as ideals with extra variables. Hence, this slightly more general approach, can be reduced to our concept of Boolean polynomials which does not suffer from the complexity of theory and implementations of Gröbner bases over a ring with zero divisors.

A big motivation for this work was the observation, that a specialized implementation can often tackle much harder problems. For instance, see Faugère's attacks on hidden field equation (HFE) cryptosystems [2003]. In this way, we wanted to provide a framework for computations with Boolean polynomials. Moreover, the shear amount of signals in many digital circuits motivated the ability to deal with a very large number of variables.

Although the treatment of polynomial systems using Gröbner bases improved considerably in recent years, current implementations have not been capable of satisfactorily handling Boolean polynomials from real-world applications yet.

Most ideas of this thesis were implemented in the PolyBoRi framework which was developed together with Alexander Dreyer [Brickenstein and Dreyer, 2009a]. It is very clear that tackling these huge systems is only possible if effective optimization takes place on all levels which we describe in the following:

- Theory: Chapter 2 elaborates the very special case of Boolean polynomials. We have formulated and proven many of their basic properties. It was known for a long time that Boolean polynomials correspond directly to propositional logic [Agnarsson et al., 1984]. These basics are enriched in this thesis by new theoretical results like the function of the complete homogeneous polynomial, the linear lead criterion, Gröbner bases computations using shift symmetry and absolute factorization of Boolean polynomials. The presentation of the theory largely follows and extends the author's contributions to [Brickenstein, Dreyer, Greuel, Wedler, and Wienand, 2009] and Brickenstein and Dreyer [2009a]. The description of the theory reuses as much general Gröbner basis theory as possible: While we consider the polynomials in PolyBoRi to be the canonical representatives of residue classes in the quotient ring modulo field polynomials, our theory covers the polynomial ring itself and ideals containing all field polynomials. For this, we utilize the 1–1 correspondence of such ideals to those in the quotient ring. Hansen and Michon [2006] gave an alternative approach applying their theory directly to the quotient ring.

- Data structures: We use zero-suppressed decision diagrams [Ghasemzadeh, 2005, Minato, 1993] for storing the polynomial term structure. While we started the project without knowledge about previous approaches going in a similar direction, we found out about first attempts to store more general polynomials in decision diagrams [Minato, 1995]. Minato treated general polynomials with integer coefficients. Independent from our research and probably at about the same time, ZDDs were used for Boolean polynomials where they are much more appropriate. Chai, Gao, and Yuan [2008] were able to reduce the memory consumption of characteristic set methods using ZDDs. It is common to all these approaches that ZDDs were seen more or less as a good way to store the polynomials (with the goal to save memory). Moreover, it is worth mentioning that the same diagram structures will be generated, when storing the *polynomial function* (not the term structure) in a functional decision diagram (FDD). Functional decision diagrams have been introduced in [Kebschull et al., 1992, Kebschull and Rosenstiel, 1993]. Computational algebra however requires many operations which are beyond storage and basic arithmetic: The implementation of monomial orderings on these data structures is quite challenging: In [Brickenstein, Dreyer, Greuel, Wedler, and Wienand, 2009] we have described this for several monomial orderings for the task of leading term computation. In this thesis the much more general case of iteration of the terms in monomial ordering is presented. The corresponding ideas and algorithms belong to the scientific results of this dissertation (see chapter 3). The implementation of these data structures was done in PolyBoRi by Alexander Dreyer.

- Algorithms: PolyBoRi covers a large variety of algorithms. We started with basic arithmetic. The CUDD [Somenzi, 2005] library did not even contain the XOR-operation for ZDDs (our addition) or the and-operation for FDDs (our multiplication). The presentation of basic addition and multiplication is close to the corresponding part in [Brickenstein, Dreyer, Greuel, Wedler, and Wienand, 2009]. It is extended by a new algorithm for fast multiplication with Boolean polynomials based on similar ideas as Karatsuba's univariate polynomial multiplication [Karatsuba and Ofman, 1962]. It was generalized to the multivariate case and specialized for Boolean multiplication (multiplication in the quotient ring modulo field equations). In the language of FDDs: we have an algorithm for the and-operation that has less recursive calls than the naïve one.

We implemented many Gröbner bases algorithms, for instance symmgbGF2 which is an advanced form of slimgb. There exist variants based on matrices using the M4RI library for linear algebra over $\mathbb{Z}_2$ [Albrecht and Bard, 2008]. Other variants only use our polynomial data structures in form of ZDDs.

We have combined that with a set of very special normal form algorithms which can efficiently produce a lexicographical normal form against a system of Boolean polynomials with linear leading terms and field equations. This is highly useful for many applications in cryptanalysis and formal verification. The combination with classical, more general, algorithms can be done as a preprocessing step for the complete algorithm. Also, we apply it from time to time during the algorithm. In particular, using the full power of ZDDs these algorithms can be combined finely with linear algebra. By rewriting monomials with such systems, we were able to limit the number of columns in our matrices.

Furthermore, we implemented the FGLM algorithm [Faugère et al., 1993] in PolyBoRi for converting between different orderings. This was very tricky indeed because due to the degree bound of one per variable not all edges and vertices in the FGLM algorithm were representable as monomials in PolyBoRi. Therefore, we replaced those steps of the original algorithm by set operations on the ZDDs producing the same result.

In contrast to all other approaches the strength of our work is that we tackle problems from both sides: using algebraic methods as well as algorithms that directly operate on the decision diagram structure. One example for that are recursive normal form functions for systems with linear lexicographical lead terms. In this particular case the normal form algorithms introduced in section 4.3 can be seen as a higher level arithmetic operation: normal forms are built on top of multiplication, just as multiplication can be built on top of addition.

Another example is the interpolation algorithm in section 5.1. It produces an interpolation polynomial which is minimal with respect to lexicographical ordering. In this way, we can compute normal forms against the vanishing ideal of a given set of points without determining the actual Gröbner bases. Addition/symmetric difference, union, intersection of sets and a recursive function for calculation of zero sets form together an interpolation algorithm.

All these algorithms are recursive on the ZDD structure, fully cacheable and do not depend on the number of points/terms, but only on the structure of the decision diagram. The presentation of the interpolation problem largely follows [Brickenstein and Dreyer, 2008].

- Model/Problem formulation: Obviously the adequate problem formulation is one of the most powerful possibilities to speed up the computation. This can involve the following areas:
  - monomial orderings,
  - equation formulation,

- formulation as Gröbner basis problem or simple normal form computation,
- determination of the Boolean variables,
- splitting of the problem in separate smaller problems,
- model intrinsic aspects. For example in cryptanalysis: How many pairs of plain-/ciphertext are treated?

Section 5.2 on formal verification treats several of these aspects. Note, that for a badly formulated problem it will be hardly possible to find an algorithm which solves it in reasonable time.

We have published several articles on the topic of Boolean Gröbner bases: Bulygin and Brickenstein [2010], Brickenstein and Dreyer [2009a], Brickenstein, Dreyer, Greuel, Wedler, and Wienand [2009], Brickenstein and Dreyer [2008]. Furthermore, this work has been made publicly available to the community and been used by foreign researchers: Guerrini, Orsini, and Simonetti [2009], Gligoroski, Dimitrova, and Markovski [2009], Gao [2009] or Albrecht and Cid [2009] whereas the latter has been one of the most promising results in algebraic cryptanalysis. A more detailed overview on the performance of PolyBoRi and the results obtained using it is given in section 6.3.

I have presented the results of the thesis at a number of conferences:

- Rhine Workshop on Computer Algebra (2006, Basel),
- Special Semester on Gröbner Bases (2006, Linz),
- Sage Days 6 (2007, Bristol),
- MEGA (2007, Strobl),
- Sage Days 10 (2008, Nancy),
- ISSAC (2008, Linz)

Acknowledgements

I would like to thank everybody, who has made this work possible. I am very much conscious about the fact, that is not possible to name everybody here, as it are the small things, that matter, and every single piece of encouragement can be invaluable.

First of all, I would like to thank my supervisor Gert-Martin Greuel, who supported my work in numerous ways: Not only, he spent much of his precious time for me, but also, he believed in me from the very beginning. This trust forms an essential foundation for this work. I also thank my second supervisor Gerhard Pfister for his support of my work. Moreover my thank goes to the complete Singular team directed by them. In particular, I would like to mention Hans Schönemann, who told me all secrets of computing Gröbner bases. Furthermore, Victor Levandovskyy and Oleksandr Motsak for explained everything about Plural to me.

Many thanks goes to Alexander Dreyer, who was the ideal research partner for me in the PolyBoRi project. Not only, he is an excellent programmer and provided much of the project infrastructure, but also, he has been a good friend to me. He also invested a lot of time in proofreading of this thesis and shared his deep LaTeX knowledge with me.

Thanks goes to my cooperation partners: Markus Wedler and the working group "Entwurf informationstechnischer Systeme" for their input on formal verification. Moreover, I appreciate the hints and literature references of Dominik Stoffel regardings the complexity of FDD sizes for multipliers. I thank Stanislav Bulygin for all his efforts in our joint work on cryptographic systems. Thanks goes to Martin Albrecht and Burcin Erocal for the integration of the PolyBoRi framework in the Sage software system. Also I had much exchange with Martin about cryptographic issues.

Thank goes also for proofreading of parts of this thesis to Martin Albrecht, Stanislav Bulygin, and Thomas Halfmann.

Moreover, I'd like to thank Armin Biere for our very fruitful discussion about Gröbner bases and SAT-solvers. It was very inspiring for me and enriched my view on this domain.

Thank goes to everybody who contributed bug fixes to PolyBoRi or its Sage interface. I would like to name here: Martin Albrecht, Michael Abshoff, Burcin Erocal, Minh Nguyen, David Kirkby, and last but not least Ralf Philipp Weinmann. Furthermore, I would like to thank Tim Abott for the Debian package of PolyBoRi.

Very special thanks goes to everybody, who morally supported my work. They gave me the needed strength and cheered me up. I would like to mention here my best friend Hannah Markwig and her husband Thomas Markwig at the first place, but also my friends Lesya Bordnarchuk, Nadine Cremer, Michael Kunte, Michael Cuntz, and Eva-Maria Zimmermann. I'd like to mention the invaluable trust and support my family is giving me, in particular I'd like to thank my mother and my sister Susanne. Thanks to everybody, who believed in me, when I gave up hope, and gave me back the trust in myself.

Chapter 1

General Gröbner basis theory

While the central part of this work is focussed on the finite field Z_2, this chapter will describe the general Gröbner basis theory over any field and the "slimgb" algorithm, which form central building blocks of the advanced techniques in the Boolean domain.

1.1 Classical notions

We present in this chapter classical results and notions from computational algebra and algebraic geometry. Whenever it is appropriate, we follow the concepts in Greuel and Pfister [2002]. For this thesis we define, that the **natural numbers** begin with 0: $\mathbb{N} = \{0, 1, 2, \ldots\}$. Let $P = K[x_1, \ldots, x_n]$ be the polynomial ring over the field K. A **monomial ordering** on P, more precisely, on the set of **monomials** $\{x^\alpha = x_1^{\alpha_1} \cdot \ldots \cdot x_n^{\alpha_n} | \alpha \in \mathbb{N}^n\}$, is a total ordering "$>$" with $x^\alpha > x^\beta \Rightarrow x^{\alpha+\gamma} > x^{\beta+\gamma}$ for all $\alpha, \beta, \gamma \in \mathbb{N}^n$.

We always assume that "$>$" is global (i. e. $x_i > 1$ for $i = 1, \ldots, n$ or, equivalently, "$>$" is a well-ordering).

We generalize the notion of monomial orderings on the polynomial ring, in the following way for the module P^r: We call an element $f \in P^r$ of the form $f = x^\alpha e_j$ a **monomial** (lying in **component** j, $\operatorname{comp}(x^\alpha e_j) := j$), where e_j is the j-th unit vector of P^r. The vector $\alpha \in \mathbb{N}^n$ is called its **exponent**. A monomial ordering "$>_m$" on P^r is a total ordering based on a monomial ordering "$>$" on P fulfilling the following two conditions for all $\alpha, \beta \in \mathbb{N}^n$ and all $1 \leq i, j \leq r$:

- $x^\alpha \cdot e_i >_m x^\beta \cdot e_j \Longrightarrow x^{\alpha+\gamma} e_i >_m x^{\beta+\gamma} \cdot e_j$
- $x^\alpha > x^\beta \Longrightarrow x^\alpha e_i >_m x^\beta e_i$

In the following we will not distinguish between "$>$" with "$>_m$". We say, that a monomial $x^\alpha e_i$ **divides** $x^\beta e_j$, if and only if $i = j$ and x^α divides x^β. An expression $\lambda \cdot m$ ($\lambda \in K$, m a monomial) is called a **term** and λ its **coefficient**. An arbitrary element $f \in P^r$ is called a **polynomial** (or a vector). For a monomial m and polynomial f we denote the coefficient with that m occurs in f by $\operatorname{coeff}(m, f)$.

Let $f = \sum_{\alpha,i} c_{\alpha,i} \cdot x^\alpha e_i$ ($c_{\alpha,i} \in K$) be a polynomial. Then

$$\operatorname{supp}(f) := \{x^\alpha e_i | c_{\alpha,i} \neq 0\}$$

is called the **support** of f. The **variables** of f, $\mathrm{vars}(f)$ are defined to be those variables, which occur with nonzero exponent in some monomial in $\mathrm{supp}(f)$.

We say f **lies completely in component** j, if all monomials in the support of f lie in component j. Furthermore $\mathrm{lm}(f)$ denotes the **leading monomial** of f, i. e. the biggest monomial occurring in f w. r. t. ">" (if $f \neq 0$). The corresponding term (**leading term**) is denoted by $\mathrm{lt}(f)$ and its coefficient (**leading coefficient**) by $\mathrm{lc}(f)$. The **leading component** is defined by $\mathrm{lcomp}(f) = \mathrm{comp}(\mathrm{lm}(f))$. Moreover, we set

$$\mathrm{tail}(f) := f - \mathrm{lt}(f).$$

We define the degree of a monomial as the total degree:

$$\deg(x^\alpha \cdot e_l) = \sum_i \alpha_i.$$

Given a set of variable $x_{i_1}, \ldots x_{i_k}$, we define

$$\deg_{x_{i_1},\ldots,x_{i_k}}(x^\alpha \cdot e_l) = \sum_{j=1}^{k} \alpha_{i_j}.$$

For a polynomial $f \in P^r$ we define

$$\mathrm{ecart}(f) = \deg(f) - \deg(\mathrm{lm}(f)).$$

If $F \subset P^r$ is any subset, $\mathrm{L}(F)$ denotes the **leading ideal** of F i. e. the P-submodule of P^r generated by $\{\mathrm{lm}(f) | f \in F \backslash \{0\}\}$. The S-Polynomial of $f, g \in P^r \backslash \{0\}$ with $\mathrm{lm}(f) = x^\alpha e_i$, $\mathrm{lm}(g) = x^\beta e_i$ (same leading component) is denoted by

$$\mathrm{spoly}(f, g) := x^{\gamma-\alpha} f - \frac{\mathrm{lc}(f)}{\mathrm{lc}(g)} x^{\gamma-\beta} g,$$

where $\gamma = \mathrm{lcm}(\alpha, \beta) := (\max(\alpha_1, \beta_1), \ldots, \max(\alpha_n, \beta_n))$. Moreover for two monomials in the same module component $t = x^\alpha \cdot e_i$, $u = x^\beta \cdot e_i$ we define their **least common multiple** $\mathrm{lcm}(t, u) := x^{\mathrm{lcm}(\alpha,\beta)} \cdot e_i$. This defines the least common multiple for monomials in modules. For the usual case of unique factorization domains we use the classical notion of the **least common multiple** (lcm) and **greatest common divisor** (gcd).

Note that for $r = 1$ we get the classical notions of leading monomial, S-polynomial, etc. in the polynomial ring P itself. A monomial m is called a **standard monomial** for an ideal or module I if $m \notin \mathrm{L}(I)$.

Recall that a finite set $G \subset P^r$ is called a **Gröbner basis** of an ideal (i. e. submodule) $I \subset P^r$ if $\{\mathrm{lm}(g) | g \in G \backslash \{0\}\}$ generates the P-module $\mathrm{L}(I)$ and $G \subset I$. G is simply called a Gröbner basis if it is a Gröbner basis of $\langle G \rangle_P$. It is called **minimal** if for each $g \in G$ there exists no $g \neq f \in G$ with $\mathrm{lm}(f)$ divides $\mathrm{lm}(g)$. If moreover for all elements $g \in G$ we have that $\mathrm{lc}(g) = 1$ and $\mathrm{supp}(\mathrm{tail}(g)) \cap \mathrm{L}(G) = \emptyset$, then G is called a **reduced Gröbner basis**. Given a monomial ordering, the reduced Gröbner basis of an ideal is unique [Greuel and Pfister, 2002].

Next, we state the definition of the standard representation as formulated in [Buchberger, 1985].

Definition 1.1.1 (Standard representation). *Let $f, g_1, \dots g_m \in P^r$, and let $h_1, \dots, h_m \in P$. Then*

$$f = \sum_{i=1}^{m} h_i \cdot g_i$$

is called a standard representation of f with respect to $g_1, \dots, g_m$, if

$$\forall i : h_i \cdot g_i = 0 \text{ or otherwise } \operatorname{lm}(h_i \cdot g_i) \leq \operatorname{lm}(f)$$

Definition 1.1.2. *A polynomial f is said to be reduced against a finite set of polynomials G if and only if $\operatorname{supp}(f) \cap \operatorname{L}(G) = \emptyset$. A polynomial r is called the* ***reduced normal form*** *of f against G if and only if it is reduced against G and $f - r$ has standard representation with respect to G.*

Remark 1.1.3. *As the reduced normal form against a Gröbner basis is unique, we may denote this by $r = \operatorname{REDNF}(f, G)$. Moreover, the reduced normal form does not depend on the Gröbner basis G itself, but only on the ideal spanned by G and the chosen monomial ordering. The normal form gives us an algorithmic solution to the ideal membership problem as for a Gröbner basis G we have that $\operatorname{REDNF}(f, G) = 0$ if and only if $f \in \langle G \rangle$ [Greuel and Pfister, 2002].*

The most important normal form algorithm is the normal form, as originally given by Bruno Buchberger, which is presented in algorithm 1.

Algorithm 1 nf_buchberger: Buchberger normal form algorithm

Input: $f \in K[x_1, \dots, x_n]^r$, G finite set of nonzero vectors in $K[x_1, \dots, x_n]^r$, $>$ a monomial ordering
Output: a (not necessarily reduced) normal form of f with respect to G
  if $f = 0$ **then**
    return 0
  $H := \{g \in G | \operatorname{lm}(g) \text{ divides } \operatorname{lm}(f)\}$
  if $H \neq \emptyset$ **then**
    choose $h \in H$
    return nf_buchberger(spoly(f, h), G)
  return f

Using an arbitrary normal form algorithm and a global ordering, we can construct an algorithm for reduced normal forms. This is shown in algorithm 2.

Algorithm 2 reduce: Reduced normal form algorithm

Input: $f \in K[x_1, \dots, x_n]^r$, G finite set of nonzero vectors in $K[x_1, \dots, x_n]^r$, nf a normal form algorithm, $>$ a monomial ordering
  $f := \operatorname{nf}(f, G)$
  if $f = 0$ **then**
    return 0
  return $\operatorname{lt}(f) + \operatorname{reduce}(\operatorname{tail}(f), G, '>', \operatorname{nf})$

Using the concept of normal form, we can present the classical Buchberger algorithm for computing a Gröbner basis. Most modern algorithms like F4 and "slimgb" can be seen as variants of the Buchberger algorithm. Moreover many classical algorithms like the Gaussian algorithm or the Euclidean algorithm for univariate polynomials are special cases of the Buchberger algorithm. The algorithm was presented the first time in [Buchberger, 1965].

Algorithm 3 Buchberger algorithm, calculates a Gröbner basis of F

Input: F finite set of polynomials (from $K[x_1, \ldots, x_n]^r$), nf a normal form algorithm, $>$ a monomial ordering.

```
P := {(f, g) | f, g ∈ F}
while P ≠ ∅ do
  choose (f, g) ∈ P
  P := P \ (f, g)
  r := nf(spoly(f, g), F)
  if 0 ≠ r then
    P := P ∪ {(r, f) | f ∈ F and lcomp(r) = lcomp(f)}
    F := F ∪ {r}
return F
```

The correctness of algorithm 3 is ensured by theorem 1.1.4.

Theorem 1.1.4 (Buchberger's criterion). *Let $F \subset P^r$ be a set of polynomials. If for all $f, g \in F$ with* $\mathrm{lcomp}(f) = \mathrm{lcomp}(g)$, $\mathrm{REDNF}(\mathrm{spoly}(f, g), F) = 0$, *then F is Gröbner basis of the submodule $\langle F \rangle_P$.*

Definition 1.1.5. *Given a set of polynomials $F \in K[x_1, \ldots, x_n]$, we define the **variety** of F:*

$$\mathrm{V}(F) := \{x \in K^n | f(n) = 0 \, \forall \, f \in F\}$$

*The **vanishing ideal** of a set $V \subseteq K^n$ is defined as*

$$\mathrm{I}(V) := \{f \in K[x_1, \ldots, x_n] | f(v) = 0 \, \forall \, v \in V\}.$$

Theorem 1.1.6 (Hilbert's Nullstellensatz). *Let K be an algebraically closed field, $I \subseteq K[x_1, \ldots, x_n]$ an ideal, then the following equality holds:*

$$\mathrm{I}(\mathrm{V}(I)) = \sqrt{I}.$$

The classical product criterion of Buchberger [Buchberger, 1985] for modules reads as follows:

Lemma 1.1.7 (Product criterion). *Let $f, g \in K[x_1, \ldots, x_n]^r$ lying entirely in the same module component, i. e. $f = F \cdot e_i$, $g = G \cdot e_i$ for some F, $G \in P$. If* $\mathrm{lm}(f) \cdot \mathrm{lm}(g) = \mathrm{lcm}(\mathrm{lm}(f), \mathrm{lm}(g))$, *then* $\mathrm{spoly}(f, g)$ *has a standard representation w. r. t. $\{f, g\}$.*

Since different monomial orderings have different important properties in Gröbner bases theory, we repeat a few basic properties and definitions here:

Definition 1.1.8. *The **lexicographical** monomial ordering is given by*

$$x^\alpha >_{lex} x^\beta \iff \exists i_0 \in \{1,\ldots,n\} : \forall i < i_0 : \alpha_i = \beta_i \text{ and } \alpha_{i_0} > \beta_{i_0}$$

*The **degree lexicographical** monomial ordering is based on the lexicographical ordering:*

$$x^\alpha >_{dlex} x^\beta \iff \deg(x^\alpha) > \deg(x^\beta) \text{ or } \deg(x^\alpha) = \deg(x^\beta) \text{ and } x^\alpha >_{lex} x^\beta.$$

*The **degree reverse lexicographical** monomial ordering is defined in the following way:*

$$x^\alpha >_{dp} x^\beta \iff \deg(x^\alpha) > \deg(x^\beta) \text{ or } (\deg(x^\alpha) = \deg(x^\beta) \text{ and } \exists i_0 : \forall i > i_0 : \alpha_i = \beta_i \text{ and } \alpha_{i_0} < \beta_{i_0}).$$

*The **ascending degree reverse lexicographical** monomial ordering is defined like the degree reverse lexicographical ordering, but in ascending order of variables ($x_1 < x_2 < x_3 \ldots$):*

$$x^\alpha >_{dpasc} x^\beta \iff \deg(x^\alpha) > \deg(x^\beta) \text{ or } (\deg(x^\alpha) = \deg(x^\beta) \text{ and } \exists i_0 : \forall i < i_0 : \alpha_i = \beta_i \text{ and } \alpha_{i_0} < \beta_{i_0}).$$

The ascending variant of the degree reverse lexicographical ordering is implemented in the POLYBORI *framework as it is more compatible with the data structures in* POLYBORI.

Theorem 1.1.9. *For every monomial ordering $>$ there exists a matrix A with n columns (as many as variables) and m rows ($m \leq n$) and entries in $\mathbb{R}$ such that A is injective as a mapping from $\mathbb{Q}^n$ to $\mathbb{R}^m$ and for all monomials t, u we have that:*

$$t > u \iff A \cdot t >_{lex} A \cdot u.$$

This theorem is due to [Robbiano, 1985]

Definition 1.1.10 (Elimination orderings). *Let $R = K[x_1,\ldots,x_n,y_1,\ldots y_m]$. An ordering $>$ is called an elimination ordering of $x_1,\ldots,x_n$ if $x_i > t$ for every monomial t in $K[y_1,\ldots,y_m]$ and every $i = 1,\ldots,n$.*

Definition 1.1.11. *Given the polynomial ring $K[x_1,\ldots,x_n]$ a **block ordering** or **product ordering** $>$ with two blocks is given by an integer $1 < i_0 \leq n$ and monomial orderings $>_1$ on $K[x_1,\ldots,x_{i_0-1}]$ and $>_2$ on $\mathbb{K}[x_{i_0},\ldots,x_n]$. We define an ordering to be a block ordering composed of $>_1$, $>_2$ if for all monomials $m_1, m_2 \in K[x_1,\ldots,x_{i_0-1}], n_1, n_2 \in \mathbb{K}[x_{i_0},\ldots,x_n]$ we have:*

$$m_1 \cdot m_2 > m_2 \cdot n_2 \iff m_1 >_1 m_2 \text{ or } (m_1 = m_2 \text{ and } n_1 >_2 n_2).$$

Ordering consisting of arbitrary many blocks are constructed recursively.

Remark 1.1.12. *Every block ordering is an elimination ordering for the variables of the first block. However not every elimination ordering is a block ordering. Example:*

$$A := \begin{pmatrix} 1 & 1 & 1 & 0 & 0 \\ 0 & 0 & 0 & 1 & 1 \\ 0 & 0 & -1 & 0 & 0 \\ 0 & -1 & 0 & 0 & 0 \\ 0 & 0 & 0 & 0 & -1 \end{pmatrix}$$

The ordering given by A ($>_A$) forms an elimination ordering on $\mathbb{Q}[x_1, x_2, x_3, y_1, y_2]$. We observe that it restricts to degree reverse lexicographical ordering on $\mathbb{Q}[x_1, x_2, x_3]$ as well as $\mathbb{Q}[y_1, y_2]$. However $>_A$ is no product ordering:

$$x_2 \cdot x_3 \cdot y_1 \cdot y_2 >_A x_1 \cdot x_2 \cdot y_1,$$

but

$$x_2 \cdot x_3 < x_1 \cdot x_2$$

As comparison we give a matrix description for a true block ordering on these variable blocks:

$$B := \begin{pmatrix} 1 & 1 & 1 & 0 & 0 \\ 0 & 0 & -1 & 0 & 0 \\ 0 & -1 & 0 & 0 & 0 \\ 0 & 0 & 0 & 1 & 1 \\ 0 & 0 & 0 & 0 & -1 \end{pmatrix}$$

Furthermore we have $\mathrm{ecart}_{>_A}(f) \leq \mathrm{ecart}_{>_B}(f)$ *for each polynomial* $0 \neq f \in \mathbb{Q}[x_1, x_2, x_3, y_1, y_2]$.

1.2 T-Representations and extended product criterion

There is an alternative approach to standard representations formulated in [Becker and Weispfennig, 1993] and utilized in [Faugère, 1999] which uses the notion of t-representations. While this notion is mostly equivalent to using syzygies, it makes the correctness of the algorithms easier to understand.

Definition 1.2.1 (t-representation). *Let t be a monomial, $f, g_1, \dots g_m \in P^r$, $h_1, \dots, h_m \in P$. Then*

$$f = \sum_{i=1}^{m} h_i \cdot g_i \in P^r$$

is called a t-representation of f with respect to $g_1, \dots, g_m$ if

$$\forall i : lm(h_i \cdot g_i) \leq t \text{ or } h_i \cdot g_i = 0.$$

Remark 1.2.2. *Note that each standard representation of f is a* $\mathrm{lm}(f)$*-representation, but t-representations are more flexible. For example, let the monomials of P be lexicographically ordered ($x > y$) and let*

$$t = x^5 y^5, g_1 = x^2, g_2 = x^5 - y, f = y.$$

Then $f = x^3 g_1 - g_2$ is a x^5y^5-representation for f. For $t < lm(f)$ t-representations of f do not exist.

Definition 1.2.3 (nontrivial t-representation)**.** *Given a representation* $p = \sum_{i=1}^{m} h_i \cdot f_i$ *w. r. t. a family of polynomials* $f_1, \dots f_m$*, then this is a t-representation for* $t = \max\{lm(h_i \cdot f_i) | h_i \cdot f_i \neq 0\}$*. We consider it trivial as p was given to us using this representation. We may shortly say that p has a* **nontrivial t-representation (w. r. t. this given presentation)***, if a t-representation of p exists with*

$$t < \max\{lm(h_i \cdot f_i) | h_i \cdot f_i \neq 0\}.$$

For example $\operatorname{spoly}(f_i, f_j)$ *has a nontrivial t-representation if there exists a representation of* $\operatorname{spoly}(f_i, f_j)$ *where the summands have leading terms smaller than*

$$\operatorname{lcm}(\operatorname{lm}(f_i), \operatorname{lm}(f_j)).$$

Theorem 1.2.4 (Extended product criterion)**.** *Let* $f = h \cdot f'$*,* $g = h \cdot g'$ *and* $f, g, f', g' \in P^r$*,* $h \in P$ *be polynomials lying completely in the same module component. If* $\operatorname{lm}(f') \cdot \operatorname{lm}(g') = \operatorname{lcm}(\operatorname{lm}(f'), \operatorname{lm}(g'))$ *then* $\operatorname{spoly}(f, g)$ *has a standard representation w. r. t.* $\{f, g\}$*.*

Proof. As the product criterion holds for f', g', there exist polynomials λ, μ, s. th. $\operatorname{spoly}(f', g') = \lambda f' + \mu g'$ is a standard representation. Then also $\operatorname{spoly}(f, g) = \lambda f + \mu g$ is a standard representation. □

Remark 1.2.5. *In practice the extended product criterion will be applied only to the case where the common factor h is a monomial. In this case it is very fast to check: for each polynomial f one computes* m_f*, the monomial* gcd *of its terms. Considering the pair f, g one only has to take the* gcd *of the two monomials* m_f*,* m_g*. However, in certain situations (e. g. cryptographic systems) one can easily detect some other factors. The speedup can be significant. This criterion, though simple, seems to have been overlooked so far.*

The following is a reformulation of the classical Buchberger's chain criterion for t-representations.

Theorem 1.2.6 (t-chain criterion)**.** *Let G be a list of polynomials, and let* $f_1, f_2, \dots, f_m \in G$ *with leading monomials involving the same module component* $\operatorname{lm}(f_i) = x^{\alpha_i} e_j$ *such that*

$$\forall i : x^{\alpha_i} | \operatorname{lcm}(x^{\alpha_1}, x^{\alpha_m}).$$

Assume that for all $i \in \{1, \dots, m-1\}$ *one of the two following conditions holds:*

1. $\operatorname{spoly}(f_i, f_{i+1})$ *has a (nontrivial) t-representation for t equal to some monomial* m_i *with* $m_i < \operatorname{lcm}(\operatorname{lm}(f_i), \operatorname{lm}(f_{i+1}))$ *with respect to G.*

2. $x^{\alpha_i} \cdot x^{\alpha_{i+1}} | \operatorname{lcm}(x^{\alpha_1}, x^{\alpha_m})$ *and* f_i*,* f_{i+1} *lie completely in the same component.*

Then $\operatorname{spoly}(f_1, f_m)$ *has a (nontrivial) t-representation for some monomial t with* $t < \operatorname{lcm}(\operatorname{lm}(f_1), \operatorname{lm}(f_m))$ *with respect to G.*

This generalizes results given in [Becker and Weispfennig, 1993], the trivial syzygy approach is inspired by [Faugère, 2002].

Proof. Without loss of generality we may assume that all f_i are normalized. Then, by assumption

$$\begin{aligned}&\text{spoly}(f_1, f_m)\\ &= t_1 \cdot \text{spoly}(f_1, f_2) + t_2 \cdot \text{spoly}(f_2, f_3) + \ldots + t_{m-1} \cdot \text{spoly}(f_{m-1}, f_m)\end{aligned}$$

where

$$t_i = \frac{\text{lcm}(x^{\alpha_1}, x^{\alpha_m})}{\text{lcm}(x^{\alpha_i}, x^{\alpha_{i+1}})}.$$

Therefore, it remains to show that each summand has nontrivial t-representation (precisely some $t < \text{lcm}(\text{lm}(f_1), \text{lm}(f_m))$).

If condition (1) holds, then the multiplication of the representation given in (1) by the monomial t_i yields a suitable t which satisfies our condition.

In the second case, one has the trivial syzygy $p_i \cdot f_{i+1} - p_{i+1} \cdot f_i = 0$ (using $f_i = p_i \cdot e_j$, $f_{i+1} = p_{i+1} \cdot e_j$). Therefore,

$$\gcd(x^{\alpha_i}, x^{\alpha_{i+1}}) \cdot \text{spoly}(f_i, f_{i+1}) = \text{tail}(p_i) \cdot f_{i+1} - \text{tail}(p_{i+1}) \cdot f_i \qquad (*)$$

forms a nontrivial t-representation. If we multiply both sides of (*) with $\frac{\text{lcm}(x^{\alpha_1}, x^{\alpha_m})}{x^{\alpha_i} \cdot x^{\alpha_{i+1}}}$ we achieve a nontrivial representation for $t_i \cdot \text{spoly}(f_i, f_{i+1})$ and some $t < \text{lcm}(x^{\alpha_1}, x^{\alpha_m})$. □

1.3 The slimgb algorithm

In this section we describe the new algorithm "slimgb" for calculations of Gröbner bases. Gröbner basis computations have made considerable progress in the sense of better data structures, highly optimized implementations, good critical pair criteria[1], algorithms like the Gröbner walk and the Hilbert-driven Buchberger, or the linear algebra approach of Faugère.

While these improvements are unable to lower the theoretical worst case complexity bounds, it has been shown very clearly that it is indeed possible to calculate the Gröbner basis in reasonable time for much more systems than one would expect. Calculating a Gröbner basis usually differs from other disciplines like for example dense linear algebra over finite fields by the crucial fact that there exists an incredible intermediate expression swell. There exists a big variety of examples where the input system is quite small, but the memory usage blows up vastly during computation while the result is just the one-polynomial. On the other hand, it has been shown that even the fastest implementations and a good selection strategy might be outperformed by a pure Buchberger algorithm in other examples. Indeed, so far it is not understood where the intermediate expression swell comes from and how to avoid it. However, in many practical cases this expression swell is unnecessary and does not appear for a different strategy. This is the motivation for "slimgb" since it will tackle the key problem, the intermediate expression swell. We will introduce a measure for intermediate expression swell - namely weighted length functions - and present an algorithm, that puts its focus directly on this fundamental problem. Rather than optimizing single operations, some extra work is done calculating these weighted length functions.

[1]The term critical pair refers to each pair still in the pair list of the Buchberger algorithm. Some of these pairs might also reduce to zero. Hence, it is possible that a pair is critical and useless at the same time.

1.3.1 The main algorithm

The main algorithm can be seen as classical Buchberger algorithm with the new reduction (or normal form) slimgbReduce.

We will use the following notation: If F is a finite tuple of objects, $F[i]$ denotes the i-th entry of F, $\#F$ denotes its length and for two tuples F, G their union $F \cup G$ is defined to be the union as sets.

Algorithm 4 slimgb main procedure, calculates a Gröbner basis of F

Input: F finite tuple of polynomials (from $K[x_1, \ldots, x_n]^r$).
  $P := \{(i,j) | 1 \leq i < j \leq \#F$ and $\mathrm{lcomp}(F[i]) = \mathrm{lcomp}(F[j])\}$ /* P is the set of critical pair indices */
  apply criteria to P
  while $P \neq \emptyset$ **do**
    choose $\emptyset \neq S \subset P$
    $P := P \backslash S$
    $(R, F) :=$ slimgbReduce(S, F) /* slimgbReduce is specified later */
    for $0 \neq r \in R$ **do**
      $F :=$ append(F, r)
      $P := P \cup \{(i, \#F) | 1 \leq i < \#F$ and $\mathrm{lcomp}(F[i]) = \mathrm{lcomp}(F[\#F])\}$
  return F

Remark 1.3.1. *When choosing critical pairs from P, one should of course take only pairs with minimal sugar value (see [Giovini et al., 1991]). One is free to use any kind of criteria to omit useless pairs (pairs reducing to zero) as long as the criteria are compatible, i. e. the question whether a pair is critical depends only on the tuple* $(\mathrm{lm}(F[1]), \ldots, \mathrm{lm}(F[\#F]))$.

For example, the chain criterion and the classical product criterion are compatible. Also Hilbert-driven Buchberger techniques belong to this category and are compatible.

In particular, the approach in this section is in some sense orthogonal to the major advancements done in [Caboara et al., 2004]. So a combination of these techniques is recommended.

The extended product criterion is also admitted as it sorts out pairs which have for a temporary state of the list F nontrivial t-representation. This property is never lost if the following specification of the reduction algorithm is used.

1.3.2 The reduction algorithm specification

In this section we define a policy for the reduction algorithm. We will formulate a concrete algorithm slimgbReduce in 1.3.5, but every algorithm which fulfills the policy is allowed.

Our reduction algorithm is a generalization of the Buchberger normal form:

- It processes several S-Polynomials in parallel.
- It gives back a possibly modified generating system F'. Whenever possible it may substitute generators by better ones with the same leading monomial, e. g. $x + y + z + 1$ by x.

- The second part R of the result is given by the nonzero reduction results of the given

Formally we define the following policy: The reduction algorithm takes as arguments the index set S (subset of the pair set) and a finite tuple of polynomials F and gives back a new set R and a finite tuple of nonzero polynomials F', s. th. the following holds:

1. $\langle F' \cup R\rangle_{K[x_1,\dots,x_n]}=\langle F\rangle_{K[x_1,\dots,x_n]}$,
2. $\#F = \#F'$,
3. F' preserves the order of F: $\forall i : \mathrm{lm}(F[i]) = \mathrm{lm}(F'[i])$ and $\mathrm{spoly}(F[i], F'[i])$ has a standard representation w. r. t. $F' \cup R$,
4. For each $(i, j) \in S$ $\mathrm{spoly}(F[i], F[j])$ has a nontrivial t-representation w. r. t. $F' \cup R$,
5. Termination: $R \neq \emptyset \Rightarrow \exists r \in R : \mathrm{lm}(r) \notin L(F)$.

Remark 1.3.2. *The third condition implies that each $f \in F$ has standard representation w. r. t. $F' \cup R$. This implies again that for each (i, j) the fact that $\mathrm{spoly}(F[i], F[j])$ has nontrivial t-representation with respect to F implies that the index pair (i, j) has the same property with respect to $F' \cup R$.*

The application of the classical Buchberger normal form of [Greuel and Pfister, 2002] for all S-Polynomials $\mathrm{spoly}(F[i], F[j])$ for $(i, j) \in S$ fulfills the desired properties of a reduction algorithm for "slimgb". Note also that using Gaussian elimination like in Faugère's F4 is a valid reduction algorithm, too. We introduce a different algorithm featuring parallel reduction modifying F by following our main strategy.

1.3.3 Termination and correctness of the main algorithm

Theorem 1.3.3. *Let $F = (f_1, \dots, f_k)$, $f_i \in K[x_1, \dots, x_n]^r$ be a polynomial system. For i, j with $\mathrm{lm}(f_i), \mathrm{lm}(f_j)$ lying in the same module component define $m_{i,j}$ as follows: If $\mathrm{lm}(f_i) = x^\alpha \cdot e_l$, $\mathrm{lm}(f_j) = x^\beta \cdot e_l$, then define $m_{i,j} = \mathrm{lcm}(x^\alpha, x^\beta)/x^\alpha$. The following are equivalent:*

1. *F is a Gröbner basis of $\langle F\rangle_{K[x_1,\dots,x_n]}$.*
2. *Set*

$$M := \{(i, j)|1 \leq i < j \leq \#F,\ \mathrm{lm}(F[i]), \mathrm{lm}(F[j]) \text{ lie in the same component}\}.$$

Then, for any subset $J \subset M$ satisfying

$$\langle\{m_{i,j} \cdot e_i|(i, j) \in J\}\rangle_{K[x_1,\dots,x_n]} = \langle\{m_{i,j} \cdot e_i|(i, j) \in M\rangle_{K[x_1,\dots,x_n]}$$

in $K[x_1, \dots, x_n]^r$, $\mathrm{spoly}(f_i, f_j)$ has a nontrivial t-representation w. r. t. F (i. e. for some $t < \mathrm{lcm}(\mathrm{lm}(f_i), \mathrm{lm}(f_j))$) $\forall (i, j) \in J$.

Proof. The proof is analogous to the proof in [Greuel and Pfister, 2002, p. 142] and therefore omitted. □

Theorem 1.3.4 (Termination and correctness). *Algorithm 4 terminates and is correct.*

Proof. In each loop of the iteration either the reduced R is empty (then the set of critical pairs becomes smaller), or there exists an f in R with $\operatorname{lm}(f) \notin \mathrm{L}(F)$. Hence the algorithm terminates if the reduction algorithm is correct and terminates.

By the properties of the reduction algorithm, $\langle F \rangle$ remains constant. Each critical index pair has a nontrivial t-representation at a certain point. By remark 1.3.2 this property is preserved.

Concerning compatibility with changing the elements of F note that each $\operatorname{lm}(F[i])$ is constant at every time. If $\operatorname{spoly}(F[i], F[j])$ has a nontrivial t-representation at some point, this property will be preserved in the future. So together with our compatibility condition for the criteria it follows that every critical pair has a nontrivial t-representation in the end. Therefore, the set F is a Gröbner basis of the module generated by the original F (the input) when the algorithm terminates. This finishes the proof of correctness for the algorithm. □

1.3.4 The strategy component: weighted lengths

One key feature of "slimgb" is the flexible, problem-oriented choice of a good weighted length algorithm. By specifying a weighted length, the algorithm gains a good selection criterion (e. g. which polynomial to use for reduction, exchange elements of F by better ones, selection of critical pairs, pivoting) which turns out to be appropriate to avoid intermediate expression swell. Note that we propose different weighted length functions depending on the ground field and on the monomial ordering.

We introduce now the notion of a weighted length by requiring certain properties which seem to be reasonable.

Definition 1.3.5 (weighted length). *A weighted length function is a map* $\operatorname{wlen} : K[x_1, \dots, x_n]^r \to \mathbb{N}$ *with the following properties:*

- $\operatorname{wlen}(p) = 0 \Leftrightarrow p = 0$
- $\operatorname{wlen}(p) \geq \#\operatorname{supp}(p)$
- *If t is a term,* $\operatorname{lm}(t) \notin \operatorname{supp}(p)$ *and* $\operatorname{lm}(p) > \operatorname{lm}(t)$ *then* $\operatorname{wlen}(p + t) > \operatorname{wlen}(p)$
- *m a monomial* $\Rightarrow \operatorname{wlen}(p) = \operatorname{wlen}(m \cdot p)$

However, none of these properties is required for the algorithm to be correct, but they are essential for the performance. Maybe in some settings one has to drop certain properties. For example, when extending this algorithm to the noncommutative case, the wlen might be nonconstant under multiplication with a monomial (as this is already the case with the normal length).

Important weighted length functions are:

- length: $\operatorname{wlen}(p) = \#\operatorname{supp}(p)$
- coefficient strategy length:

$$\operatorname{wlen}(p) = \operatorname{coeffSize}(\operatorname{lc}(p)) \cdot \#\operatorname{supp}(p)$$

(over function fields this length is usually very near to normal length if you consider the parameters as variables)

- extreme coefficient strategy length:

$$\mathrm{wlen}(p) = \mathrm{coeffSize}(\mathrm{lc}(p))^2 \cdot \#\,\mathrm{supp}(p)$$

- elimination strategy length:

$$\mathrm{wlen}(p) = \sum_{m \in \mathrm{supp}(p)} (1 + |\deg(m) - \deg(\mathrm{lm}(p))|^+)$$

(good for elimination orderings $|x|^+ := x$ if x is positive, 0 else).

- elimination and coefficient strategy length:

$$\mathrm{wlen}(p) = \mathrm{coeffSize}(\mathrm{lc}(p)) \cdot \sum_{m \in \mathrm{supp}(p)} (1 + |\deg(m) - \deg(\mathrm{lm}(p))|^+)$$

Remark 1.3.6. *When using buckets (see [Yan, 1998]) as polynomial data structure, the length can only be estimated (but this is not a serious problem). The elimination strategy weighted length uses the algebraic structure, it cannot easily be transfered to linear algebra like in Faugère's F4. The elimination strategy speeds up many lexicographical examples by up to a factor of 1000 (compared to the same algorithm with normal lengths) or even more. In our experiments the extreme coefficient strategy length was only useful in a small set of examples, but it shows that there exists room for variations.*

1.3.5 The reduction algorithm of slimgb

The idea of the reduction algorithm is to eliminate effects of disadvantageous order of computations. For this aim the reduction is done in parallel, and elements of the considered system F are exchanged by better ones which are found later.

The reduction algorithm works similar to Gaussian elimination with extra operations. First we assign

$$R := \{\mathrm{spoly}(F[i], F[j]) | (i, j) \in S\}.$$

We say that an operation takes place at a monomial m if m is the highest leading monomial of the polynomials that is involved in the operation. We define:

$$R_m := \{r \in R | \,\mathrm{lm}(r) = m\}$$

We use an operation on the system R which takes place at the highest leading monomial m for which such an operation is possible (analogous to Gaussian elimination). After each operation, we filter out zero entries from R. This is iterated as long as possible.

The allowed operations are:

1. $f \in F, \mathrm{lm}(f)$ divides m, then $R := R \backslash R_m \cup \{\mathrm{spoly}(r, f) | r \in R_m\}$
2. $r \in R, \#R_m \geq 2$, then $R := R \backslash R_m \cup \{r\} \cup \{\mathrm{spoly}(r, f) | f \in R_m \backslash \{r\}\}$

3. $r \in R$, $f = F[i]$, $\mathrm{lm}(f) = \mathrm{lm}(r)$, $\mathrm{wlen}(f) > \mathrm{wlen}(r)$, then $F[i] := r$ and $R := R \backslash \{r\} \cup \{\mathrm{spoly}(r, f)\}$

At the end the new F and R are returned.

Remark 1.3.7. *Sometimes there exists a choice whether to use the first or the second operation. This is decided by the "best" (i. e. with the lowest weighted lengths) polynomial one finds with leading monomial dividing m. If it is contained in F, the first operation is used, otherwise the second.*

Proof. Termination is clear since R is finite and the leading monomials drop w. r. t. "$>$" (well-ordering).

Correctness: Note that all properties of a valid reduction algorithm (exception: the termination property) hold at the beginning of the reduction algorithm and in each step this property is preserved. If no operation is possible any more, then either R is empty, or R is nonempty and operation (1) is not possible any more. Hence there exist $r \in R$ s. th. $\mathrm{lm}(r) \notin \mathrm{L}(F)$. □

1.3.6 Notes on the implementation of the algorithm

- Generalizing the second operation in the reduction algorithm by reducing some element f by an element r, if $\mathrm{lm}(r)$ divides $\mathrm{lm}(f)$ (instead of equality), appears not to be a good idea. In elimination examples the performance drops by factor 1000 and more. One plausible explanation is: Assuming the sugar value of f is equal to the sugar value of r, then the sugar value of $\mathrm{spoly}(f, r)$ is higher than the sugar value of f if $\mathrm{lm}(r)$ only divides $\mathrm{lm}(f)$.

- Although no data structures are prescribed, geobuckets (see [Yan, 1998]) are recommended during the reduction algorithm for the elements of R (except for the very special case of the PolyBoRi library). The elements stay in the form of geobuckets until they are used to reduce something (operation 2). Then they are brought into a canonical form, used for reduction and converted back to geobuckets. For calculating the length of a polynomial the bucket structure should not be dissolved. Either one estimates the length during generation by the formula

$$\mathrm{length}(\mathrm{spoly}(f, g)) \leq \mathrm{length}(f) + \mathrm{length}(g) - 2$$

  or one adds up the lengths of the buckets.

- If the coefficient field is $\mathbb{Q}$ or some function field, denominators should be avoided. In particular, in our experiments we assume that all coefficients lie in $\mathbb{Z}$ or in the polynomial ring in case of function fields. These are only rings and therefore content plays a role. Whenever adding an element to F or using an element from R to reduce others, the content should be extracted.

1.3.7 Noncommutative algebras

While it certainly exceeds the scope of this thesis to repeat the noncommutative Gröbner basis theory, it is worth mentioning that the algorithm presented above can be translated easily to G-algebras (for background about G-algebras see Levandovskyy [2005]) by paying attention to the following points:

- Make sure that multiplication with monomials when building S-Polynomials is always done from the left (resp. right) side; the algorithm will then return a left (resp. right) Gröbner basis.
- The normal product criterion and the extended product criterion do not hold in general noncommutive settings.

1.4 Benchmarks

1.4.1 The algorithms

This section compares "slimgb" and "std". While "slimgb" is the implementation of the algorithm in SINGULAR presented in this chapter, "std" is the standard Buchberger algorithm in SINGULAR. As "slimgb" and "std" share both the same data structures and many other functions, they are comparable w. r. t. the quality of the implementation, even if "std" is to some degree more optimized.

It is clear that our algorithm has the same worst-case complexity as Buchberger's algorithm. Therefore we can only report on practical efficiency by comparing implementations based on the same environment, in particular those based on the same data structures. Moreover, as there is no reduction strategy, which is uniformly preferred on all examples, we present a variety of different examples, most of them chosen from Symbolic Data [Gräbe, 2000-2006]. We concentrate on situations where extreme expression swell occurs during the computation (e. g. computations over function fields, the rational numbers or elimination orderings). At the moment "slimgb" uses weighted length for coefficient strategy over function fields and over the rationals in the case of homogenous systems or degree orderings. If the field is $\mathbb{Z}/p$ and if the ordering is an elimination ordering an elimination strategy is used. If the field is $\mathbb{Z}/p$ and the example is homogeneous or has a degree ordering the usual length is used. Over function fields and the rational numbers the following holds: If the ordering is an elimination ordering, then the combined weighted length for elimination and coefficients is used.

It should be emphasized, that there exists no Gröbner basis algorithm and no implementation, which shows uniformly the best performance on all examples. The same applies to "slimgb" which is designed to avoid intermediate expression swell and therefore, the overhead can become too large, if expression swell is only moderate. For instance, currently the implementation of "slimgb" always uses the weighted length function with the most information, in some settings one might wish to use a function which is faster to compute, but ignores certain aspects. However, in many examples which arise from practice the gain of performance is dramatic. Below we present some examples where the techniques used in "slimgb" provide the only way to finish a Gröbner basis computation.

1.4.2 The results

In the following some benchmarks are presented. The examples were mostly taken from Symbolic Data [Gräbe, 2000-2006]. The Turaev Viro examples arise from problems in topology and were provided by Simon King. They are, as well as the example from Gema Díaz, available at [Brickenstein et al., 2006]. The HFE example (hidden field equations) was provided by Stanislav Bulygin. The noncommutative examples were contributed by Viktor Levandovskyy.

The implementation of "slimgb" is generic and does not make use of any special tricks for the presented classes of examples. It remains a challenge for the future to provide specialized implementations for certain ground fields like the rational numbers. In this case modular techniques might help to improve performance even more. For the field $\mathbb{Z}/2$ the potentials of a sophisticated implementation of the algorithm presented in this chapter forms the foundation of the work presented in the following chapters.

The SINGULAR version used here is compiled from the sources of "Singular-3-0-2" which is freely downloadable. Timings marked with (*) were measured with Singular-3-0-1. The machine used for these benchmarks is a Dual Opteron with 2.2 GHz and 16 GB RAM. Note that all tested algorithms only make use of one CPU.

Some examples took too long in "std"; they were aborted manually after one week, or when the computation time (s = second, h = hour, d = day) exceeded the one of "slimgb" significantly by a factor of (at least) 2000. The std command was also stopped, when it consumed to much memory, i. e. in case it was using too much RAM (more than 16 GB).

The polynomial rings are denoted by $K(n)[m]$, where n is the number of parameters and m the number of variables. $\mathbb{F}$ denotes the field $\mathbb{Z}/32003$. The tables present computation time and memory consumption of the computations. The abbreviation Eq. stands for the number of equations of the example.

The first table presents examples over function fields in lexicographical ordering.

example	ring	eq.	slimgb		std	
G./Chou 274 2	$\mathbb{F}(4)[7]$	7	7.39s	1.7MB	> 228982s	> 1GB
G./Chou 302 1	$\mathbb{F}(5)[8]$	8	0.48s	1.2MB	> 150634s*	> 2700MB*
G./Chou 303 1	$\mathbb{F}(5)[8]$	8	1.15s	2.2MB	> 158370s*	> 2200MB*
G./Brahmagupta	$\mathbb{F}(4)[5]$	5	0.03s	0.7MB	15.21s	3.3MB

The second table presents examples in lexicographical ordering.

example	ring	eq.	slimgb		std	
Katsura 5	$\mathbb{F}[6]$	6	0.01s	0.8MB	0.88s	18.0MB
Katsura 6	$\mathbb{F}[7]$	7	0.27s	1.9MB	> 3284s	> 17000MB
Vermeer	$\mathbb{F}[5]$	4	0.27s	1.2MB	5.13s	46.8MB
Vermeer	$\mathbb{Q}[5]$	4	0.56s	3.1MB	> 93519s	> 5000MB
ZeroDim/ex 29	$\mathbb{F}[8]$	8	0.06s	0.8MB	> 4961s	> 17000MB
ZeroDim/ex 57	$\mathbb{F}[8]$	8	0.30s	3.0MB	> 1591s	> 17000MB
HFE 15	$\mathbb{Z}/2[15]$	8	39min	211MB	> 166min	> 17000MB

The third table contains examples over function fields in degree reverse lexicographical ordering.

example	ring	eq.	slimgb		std	
H./G./Simson 3	$\mathbb{F}(4)[10]$	9	26.50s	26.3MB	599s	18.3MB
H./G./Simson 3	$\mathbb{Q}(4)[10]$	9	27.59s	26.1MB	670s	136MB
G./Chou 94 1	$\mathbb{Q}(3)[7]$	7	0.21s	0.6MB	4.23s	1.6MB
G./Chou 274 2	$\mathbb{Q}(4)[7]$	7	0.13s	0.6MB	2.47s	1.6MB
G./Chou 160 1	$\mathbb{Q}(3)[6]$	6	0.03s	0.6MB	0.28s	0.6MB

The fourth table presents examples in degree ordering.

example	ring	eq.	slimgb		std	
Diaz 1	$\mathbb{Q}[7]$	9	0.276s	4.8MB	> 160000s	> 1000MB
Viro 44	$\mathbb{Q}[44]$	1661	1min	156MB	4 days	900MB
Viro 111	$\mathbb{Q}[111]$	10159	20min		> 1week*	
Viro 53	$\mathbb{Q}[53]$	892	1h		> 1week*	

Finally, the following table shows some noncommutative examples. In this area "slimgb" can handle the expression swell by far better than the Buchberger implementation "std". In this table block orderings, lexicographical (lp) and degree reverse lexicographical (dp) examples are shown.

Moreover, we would like to refer to [Andres, in preparation], where much more extensive, noncommutative benchmarks for Reiffen-curves with "slimgb" have been done. In many examples "slimgb" needed just a few seconds, while "std" was terminated because of exceeding the time limit (1 hour) or because it was out of memory.

example	ring	eq.	ord	slimgb		std	
Ucha 2	$\mathbb{Q}[7]$	3	block	3s	8MB	> 200s	> 13000MB
Ucha 4	$\mathbb{Q}[7]$	3	lp	0.015s	0.64MB	> 0.22s	18MB
Tarasov 2	$\mathbb{Q}[7]$	2	dp	15s	18MB	> 10000s*	> 2400MB*
Bernstein 5	$\mathbb{Q}[7]$	8	block	1min	148MB	> 2d*	> 7500MB*

We measured the time of many more examples, which showed that "slimgb" performs much better than "std" for elimination orderings (any field), for function fields (any ordering) and for noncommutative systems (of course without the extended product criterion) and often better for rational numbers and degree orderings.

We also checked the implementations of F4 [Faugère, 1999] in Magma 2.13-2 [Bosma et al., 1997] and Maple/FGb [Faugère, 2006]. These experiments have not been exhaustive, however, they already demonstrated achievements of the techniques presented in this article. For example, Magma did not succeed with the smallest Turaev/Viro example, we terminated the execution after 6 days with 17GB of memory (including swap memory). The 32-bit version of FGb failed after 7 minutes with a memory allocation error in the same example. This illustrates the fact that there exists no overall performant Gröbner basis algorithm.

Of course, one can compare "slimgb" also to some indirect algorithms – for example to a version of "groebner", which first homogenizes, calculates some degree reverse lexicographical Gröbner basis, then uses the Hilbert-driven Buchberger algorithm for a lexicographical Gröbner basis and finally dehomogenizes. We found when comparing in SINGULAR, for half of elimination examples "slimgb" is by far better, on half of them the other Hilbert-driven algorithm is by far better. The differences are huge in both directions. However, these are completely different algorithms with different strengths and weaknesses. So we need them both. Therefore it is important that the direct algorithms regain ground with "slimgb".

Chapter 2

Boolean polynomials and Boolean Gröbner bases

In this chapter we elaborate the theory of Boolean polynomials. We have formulated and proven many of their basic properties. It was known for a long time that Boolean polynomials correspond directly to propositional logic [Agnarsson et al., 1984]. Central new results of this chapter are:

- the function of the complete homogeneous polynomial,
- the linear lead criterion,
- Gröbner bases computations using shift symmetry,
- absolute factorization of Boolean polynomials, and
- greedy normal form and proof its better properties in case of Boolean polynomials.

We start with definitions and various properties of our special case.

2.1 Boolean polynomials

In this section we model expressions from propositional logic as polynomial equations over the finite field with two elements. Using algebraic language the problem of satisfiability can be approached by a tailored Gröbner basis computation. We start with the polynomial ring $\mathbb{Z}_2[x_1, \ldots, x_n]$.

Since the considered polynomial functions take only values from $\mathbb{Z}_2$, the condition $x = x^2$ holds for all values of x. Hence, it is reasonable to simplify a polynomial in $\mathbb{Z}_2[x_1, \ldots, x_n]$ w. r. t. the **field equations**

$$x_1^2 = x_1,\ x_2^2 = x_2, \quad \ldots \quad ,\ x_n^2 = x_n\,. \tag{2.1}$$

We call the polynomials $x_1^2 + x_1, \ldots, x_n^2 + x_n$ **field polynomials.**

Definition 2.1.1. *A (multivariate) polynomial f in $\mathbb{Z}_2[x_1, \ldots, x_n]$, where each term has at most degree 1 per variable, is called a **Boolean polynomial**:*

$$\deg_{x_i}(f) \leq 1 \quad \forall i \in \{1, \ldots, n\}.$$

We define $\mathbb{B}$ to be the set of all Boolean polynomials. Let $v = (v_1, \ldots, v_n)$ be a vector in $\{0, 1\}$. Then we define $\mathrm{setvars}(v) = \{x_i | v_i = 1\}$, *the* ***variable set associated to*** *v.*

The following statements are not difficult to prove, but essential for the whole theory.

Theorem 2.1.2. *The composition $\mathbb{B} \hookrightarrow \mathbb{Z}_2[x_1, \ldots, x_n] \twoheadrightarrow \mathbb{Z}_2[x_1, \ldots, x_n]/\langle x_1^2+x_1, \ldots, x_n^2+x_n\rangle$ is a bijection. That is, the Boolean polynomials are a canonical system of representatives of the residue classes in the quotient ring of $\mathbb{Z}_2[x_1, \ldots, x_n]$ modulo the ideal of the field polynomials $\langle x_1^2 + x_1, \ldots, x_n^2 + x_n\rangle$. Moreover, this bijection provides $\mathbb{B}$ with the structure of a $\mathbb{Z}_2$-algebra.*

Proof. The map is certainly injective. Since any polynomial can be reduced to a Boolean polynomial using $\{x_1^2 + x_1, \ldots, x_n^2 + x_n\}$, the map is also surjective. □

Definition 2.1.3. *A function $f : \mathbb{Z}_2^n \to \mathbb{Z}_2$ is called a* ***Boolean function****. We define the* ***Boolean multiplication*** *of two Boolean polynomials p and q to be the Boolean polynomial representing the product of their residue classes. Equivalently*

$$p \star q = \mathrm{REDNF}(p \cdot q | x_1^2 + x_1, \ldots, x_n^2 + x_n).$$

Proposition 2.1.4. *Polynomials in the same residue class modulo $\langle x_1^2+x_1, \ldots, x_n^2+x_n\rangle$ generate the same function.*

Proof. Let p, q be polynomials with $p - q \in \langle x_1^2 + x_1, \ldots, x_n^2 + x_n\rangle$. By theorem 2.1.2 we have

$$p = b + f_p, q = b + f_q\,,$$

where the first summand b is a common Boolean polynomial and the second summand lies in $\langle x_1^2 + x_1, \ldots, x_n^2 + x_n\rangle$. The latter evaluates to zero at each point in $\mathbb{Z}_2^n$. □

Theorem 2.1.5. *The map from $\mathbb{B}$ to the set of Boolean functions $\{f : \mathbb{Z}_2^n \to \mathbb{Z}_2\}$ by mapping a polynomial to its polynomial function is an isomorphism of $\mathbb{Z}_2$-vector-spaces. Even more, it is an isomorphism of $\mathbb{Z}_2$-algebras.*

Proof. The map is clearly a $\mathbb{Z}_2$-algebra homomorphism. Injectivity follows from theorem 2.1.2 together with Proposition 2.1.4. For surjectivity it suffices to see that both sides have dimension 2^n. □

Corollary 2.1.6. *Every Boolean polynomial $p \neq 1$ has a zero over $\mathbb{Z}_2$. Every Boolean polynomial $p \neq 0$ has a one over $\mathbb{Z}_2$ that is $p + 1$ has a zero.*

Corollary 2.1.7. *There is a natural one-to-one correspondence between Boolean polynomials and algebraic subsets of $\mathbb{Z}_2^n$ given by $p \mapsto \mathrm{V}(\langle p, x_1^2 + x_1, \ldots, x_n^2 + x_n\rangle)$. Moreover, every subset of $\mathbb{Z}_2^n$ is algebraic.*

Proof. Since $\mathbb{Z}_2^n$ is finite, every subset is algebraic. Let χ_S be the characteristic function of a subset $S \subseteq \mathbb{Z}_2^n$ that is $\chi_S(\mathbf{x}) = 1$ if and only if $\mathbf{x} \in S$. By theorem 2.1.5 there is a $p \in \mathbb{B}$ defining $1 + \chi_S$. Hence, the map is surjective. Moreover, since both sets have the same cardinality, the results follows. □

After showing the correspondence between Boolean functions and Boolean polynomials we have a look at Boolean formulas, the kind of formulas defining Boolean functions.

Definition 2.1.8. *We define a map ϕ from formulas in propositional logic to Boolean functions by providing a translation from the basis system* `not` *($\neg$),* `or` *($\vee$),* `true` *(True). For any formulas p, q we define the following rules*

$$\begin{aligned} \phi(p \vee q) &:= \phi(p) \cdot \phi(q) \\ \phi(\neg p) &:= 1 - \phi(p) \\ \phi(\text{True}) &:= 0 \end{aligned} \tag{2.2}$$

Recursively, every formula in propositional logic can be translated into Boolean functions as $\{\vee, \neg, \text{True}\}$ forms a basis system in propositional logic (it generates the Boolean algebra).

Remark 2.1.9. *It is quite natural to identify* 0 *and* True *in computer algebra as we usually associate to a polynomial f the equation $f = 0$, and f being zero is equivalent to the equation being fulfilled.*

Theorem 2.1.10. *Let $f = p \cdot q$, $f \in \mathbb{Z}_2[x_1, \ldots, x_n]$ a polynomial with coefficients in $\mathbb{Z}_2$, $p, q \in \bar{\mathbb{Z}}_2[x_1, \ldots, x_n]$, polynomials over the algebraic closure. If f is a Boolean polynomial, then p and q have a degree of at most one in each variable. Furthermore* $\text{vars}(p) \cap \text{vars}(q) = \emptyset$.

Proof. For the first claim, we just have to show that each variable occurs with degree of at most 1 in p, q. Let $i \in \{1, \ldots, n\}$.

$$1 \geq \deg_{x_i}(f) = \deg_{x_i}(p) \cdot \deg_{x_i}(q) \Longrightarrow \deg_{x_i}(p) \leq 1, \deg_{x_i}(q) \leq 1.$$

The second claim also follows immediately from the degree formula as a variable occurring in both factors would result in a product which is at least quadratic in this variable. □

Remark 2.1.11. *If the factors p and q of theorem 2.1.10 are polynomials with coefficients in $\mathbb{Z}_2$, then they are Boolean polynomials.*

Theorem 2.1.12. *Let $f = p \cdot q$, $f \in \mathbb{Z}_2[x_1, \ldots, x_n]$, $p, q \in \bar{\mathbb{Z}}_2[x_1, \ldots, x_n]$. Then we have* $\text{length}(f) = \text{length}(p) \cdot \text{length}(q)$.

Proof.

$$f = \sum_{m \in \text{supp}(p)} \sum_{n \in \text{supp}(q)} \text{coeff}(m, p) \cdot \text{coeff}(n, q) \cdot m \cdot n$$

We show that each summand yields a term with a different exponent. Let $m_1, m_2 \in \text{supp}(p)$, $n_1, n_2 \in \text{supp}(q)$ satisfying:

$$t := m_1 \cdot n_1 = m_2 \cdot n_2.$$

Then for each variable $v \in \text{vars}(p)$, v divides t if and only if v divides m_1 (resp. m_2). Since we have no coefficients and degree bound of one per variable, it follows that m_1 and m_2 coincide. Analogously, we get the same result for n_1, n_2. Since each summands yields a different monomial, we get

$$\text{length}(f) = |\text{supp}(p)| \cdot |\text{supp}(q)| = \text{length}(p) \cdot \text{length}(q).$$

□

Remark 2.1.13. *This situation is different than in the usual polynomial ring. For example over the rational numbers we have* $(x-1)\cdot(x^2+x+1) = x^3 - 1$

Corollary 2.1.14. *A non-constant Boolean polynomial* p *is irreducible if* $\text{length}(p)$ *is a prime number and it is not divisible by variables.*

Theorem 2.1.15. *Let* f *be a Boolean polynomial. A factorization of* f *over the algebraic closure (absolute factorization) can be done using Boolean polynomials in* $\mathbb{Z}_2[x_1, \ldots, x_n]$*.*

Proof. For the case of irreducible polynomials the claim is trivial. Without loss of generality, we can assume that f splits into two irreducible factors. So, let $0 \neq f = p \cdot q$ be a nontrivial factorization (p, q no units). Hence, we can change the factorization by multiplying the first factor with a unit and the second with the inverse of that unit, we can assume that p contains a term of the form x^α.

As we have seen in theorem 2.1.12, each product of a term in p and q occurs as single term in $f = p\cdot q$. Because p contains a term with coefficient 1 all coefficients of q will occur in f. Since f is polynomial over $\mathbb{Z}_2$, it follows that q is also polynomial over $\mathbb{Z}_2$. Hence, all non-zero coefficients are one, and therefore we can apply the same argumentation to see that p is a polynomial over $\mathbb{Z}_2$. Using theorem 2.1.10 we conclude that p and q are Boolean polynomials. □

Definition 2.1.16 (division with remainder). *For Boolean polynomials* f*,* g *we define*

$$h := \frac{f}{g}$$

to be the result of ***dividing*** f *by* g*:*

$$f = h \cdot g + r \qquad \textbf{\textit{with remainder}}\ r = REDNF(f, \{g\}).$$

2.1.1 The complete homogeneous polynomial

In this section we will introduce, the complete homogeneous polynomial of degree d. We will do so for several reasons:

- It is very structured.
- It is useful in many computations.
- Its polynomial function is important for many applications.

Definition 2.1.17. *The* ***complete homogeneous polynomial of degree*** d *in* $\{x_1, \ldots, x_n\}$ *is defined to be the sum of all monomials in* $\{x_1, \ldots, x_n\}$ *of degree* d*.*

Example 2.1.18. *For variables* x*,* y*,* z *the complete homogeneous polynomial of degree* 2 *is* $x \cdot y + x \cdot z + y \cdot z$*.*

Theorem 2.1.19. *Let* p *be a prime number,* $p^k > n, m \in N$ *non-negative integers be decomposed in the following way.*

$$n = \sum_{i=0}^{k-1} p^i \cdot n_i, 0 \leq n_i < p$$

$$m = \sum_{i=0}^{k-1} p^i \cdot m_i, 0 \leq m_i < p$$

Then the following product formula holds:

$$\binom{m}{n} \equiv \prod_{i=0}^{k-1} \binom{m_i}{n_i} \mod p.$$

This is known as ***Lucas' Theorem*** *[Lucas, 1878].*

We apply this classical theorem to get an description of the polynomial function of the homogeneous polynomial of degree d.

Theorem 2.1.20. *Let $d = 2^r$ be a power of two, $V = \{x_1, \ldots, x_n\}$. The polynomial function of the complete homogeneous polynomial of degree d maps a vector with m one entries to one if and only if the r-th binary digit of m is one.*

Proof. Let $V = \{x_1, \ldots, x_n\}$ be a set of variables, $d \geq 0$, and p the complete homogeneous polynomial of degree d in V. p has $\binom{n}{d}$ terms. Let v be a vector in $\{0,1\}^n$, and $U =$ setvars(v). Let m be the number of elements in U. If all member of $V \backslash U$ are replaced by 0 in p, then we get the complete homogeneous polynomial of degree d in U. This polynomial has $\binom{m}{d}$ terms. If 1 is plugged in for these variables, then the result is of course $\binom{m}{d}$ mod 2. A binomial $\binom{a}{b}$ coefficient is 1, if $b = 0$, and a for $b = 1$. Let $m = \sum_{i=0}^{k-1} m_i \cdot 2^i$ be the binary representation of m (for some suitable $k > r$). So, by Lucas's Theorem 2.1.19

$$\binom{m}{d} = \binom{m}{2^r} \equiv \left(\prod_{i \in \{0,\ldots,r-1,r+1,\ldots,k-1\}} \binom{m_i}{0} \right) \cdot \binom{m_r}{1} = \binom{m_r}{1} = m_r$$

is congruent modulo two to the r-th binary digit m_r of m. □

In this way the complete homogeneous polynomial can be used to build a word as sum of m values in $\{0,1\}$. The r-th bit function is represented by the complete polynomial of degree 2^r. As we will see in section 3.2.4 this polynomial can be represented in a very compact form.

2.2 Boolean Gröbner Bases

In this section, we present the theory of Gröbner bases over Boolean rings. In the following, we always assume that the monomial ordering is global, i. e. $\text{lm}(x^2 + x) = x^2$ for every variable x. Since $(\mathbb{B}, +, \star) \cong \mathbb{Z}_2[x_1, \ldots, x_n]/\langle x_1^2 + x_1, \ldots, x_n^2 + x_n \rangle$, this is mathematically equivalent to the theory of Gröbner bases of ideals containing the quotient ideal. In the classical setting this would mean to add the field polynomials $x_1^2 + x_1, \ldots, x_n^2 + x_n$ to the given generators $S \subseteq \mathbb{B}$ of a polynomial ideal and compute a Gröbner basis of the ideal $\langle S, x_1^2 + x_1, \ldots, x_n^2 + x_n \rangle$ in $\mathbb{Z}_2[x_1, \ldots, x_n]$. This general approach is not well-suited for the special case of ideals representing Boolean reasoning systems. Therefore, we propose and develop algorithmic enhancements and improvements of the underlying theory of Gröbner bases for ideals over $\mathbb{Z}_2[x_1,\dots,x_n]$ containing the field equations. Using Boolean multiplication this is implementable directly via computations with canonical representatives of the residue classes in the quotient ring. The following theorem shows that it suffices to treat the Boolean polynomials introduced in definition 2.1.1 only.

Theorem 2.2.1. *Let $S \subseteq \mathbb{Z}_2[x_1,\dots,x_n]$ be a generating system of some ideal such that $\{x_1^2+x_1,\dots,x_n^2+x_n\} \subseteq S \subseteq \mathbb{B} \cup \{x_1^2+x_1,\dots,x_n^2+x_n\}$. Then all polynomials created in the classical Buchberger algorithm applied to S are either Boolean polynomials or field polynomials if a reduced normal form is used.*

Proof. All input polynomials fulfill the claim. Furthermore, every reduced normal form of an S-Polynomial is reduced against $\{x_1^2+x_1,\dots,x_n^2+x_n\}$, so it is Boolean. Moreover, using Boolean multiplication every polynomial inside the normal form algorithm is Boolean. Using Boolean multiplication at this point is equivalent to usual multiplication and a normal form computation against the ideal of field equations afterwards. □

Remark 2.2.2. *Using this theorem we need field equations only in the generating system and the pair set. On the other hand, we can implicitly assume that all field equations are in our polynomial set and then replace the pair (x_i,p) (using Boolean multiplication) by the Boolean polynomial given as $x_i \star p = \mathrm{REDNF}(\mathrm{spoly}(x_i,p)|x_1^2+x_1,\dots,x_n^2+x_n)$. In this way we can eliminate the field equations completely. A more efficient implementation would represent the pair by the tuple (i,p) as this still allows the application of the criteria, but delays the multiplication. Similar techniques can be applied for the implementation of Gröbner bases in the exterior algebra.*

Lemma 2.2.3. *The set of field equations $\{x_1^2+x_1,\dots,x_n^2+x_n\}$ is a Gröbner basis.*

Proof. Every pair of field equations has a standard representation by the product criterion. Hence $\{x_1^2+x_1,\dots,x_n^2+x_n\}$ is a Gröbner basis by Buchberger's Criterion [Greuel and Pfister, 2002, Theorem 1.7.3] □

Theorem 2.2.4. *Every $I \subseteq \mathbb{Z}_2[x_1,\dots,x_n]$ with $I \supseteq \langle x_1^2+x_1,\dots,x_n^2+x_n\rangle$ is radical.*

Proof. Consider $p \in \mathbb{Z}_2[x_1,\dots,x_n]$, w. l. o. g. assume p is reduced against the leading ideal $\mathrm{L}(I)$. In particular $\mathrm{lm}(p)$ is a Boolean polynomial. Let $r>0$ and q be the unique reduced normal form of p^r w. r. t. the field ideal. So q is also a Boolean polynomial. Since p^r-q is a linear combination of field equations, p^r-q is the zero function over $\mathbb{Z}_2$. By Corollary 2.1.6 we get $p=q$ since p^r and p define the same Boolean function. Suppose now $p^r \in I$. Then we have $p=q=p^r-(p^r-q) \in I$ since $I \supset \langle x_1^2+x_1,\dots,x_n^2+x_n\rangle$. □

Note that for $\{x_1^2+x_1,\dots,x_n^2+x_n\} \subseteq I \subseteq \mathbb{Z}_2[x_1,\dots,x_n]$ the algebraic set $\mathrm{V}(I)$ is equal to the a priori larger set $\{\mathbf{x} \in \overline{\mathbb{Z}_2}^n | f(x)=0 \forall f \in I\}$ where $\overline{\mathbb{Z}_2}$ denotes the algebraic closure of $\mathbb{Z}_2$. Hence we have

Corollary 2.2.5. *For ideals $I \subseteq \mathbb{Z}_2[x_1,\dots,x_n]$ with $I \supseteq \langle x_1^2+x_1,\dots,x_n^2+x_n\rangle$ the following stronger version of Hilbert's Nullstellensatz (theorem 1.1.6) holds:*

1. $I=\langle 1\rangle \iff \mathrm{V}(I)=\emptyset$,

2. $\mathrm{I}(\mathrm{V}(I)) = I$.

Lemma 2.2.6. *If* $I = \langle p, x_1^2 + x_1, \ldots, x_n^2 + x_n \rangle$ *then* $V(I) = V(p)$ *and every polynomial* $q \in \mathbb{Z}_2[x_1, \ldots, x_n]$ *with* $\mathrm{V}(q) \supset \mathrm{V}(p)$ *lies in* I.

Proof. Simple application of Hilbert's Nullstellensatz. □

It is an elementary fact that systems of logical expressions can be described by a single expression which describes the whole system behaviour. Hence, the one-to-one correspondence of Boolean polynomials and Boolean functions given by the mapping defined in definition 2.1.8 motivates the following theorem.

Theorem 2.2.7. *Every ideal in* $\mathbb{Z}_2[x_1, \ldots, x_n]/\langle x_1^2 + x_1, \ldots, x_n^2 + x_n \rangle$ *is generated by the equivalence class of one unique Boolean polynomial. In particular,* $\mathbb{Z}_2[x_1, \ldots, x_n]/\langle x_1^2 + x_1, \ldots, x_n^2 + x_n \rangle$ *is a principal ideal ring (but not a domain).*

Proof. We use the one-to-one correspondence of ideals in the quotient ring and ideals in $\mathbb{Z}_2[x_1, \ldots, x_n]$ containing $\langle x_1^2 + x_1, \ldots, x_n^2 + x_n \rangle$. Therefore, let $\langle x_1^2 + x_1, \ldots, x_n^2 + x_n \rangle \subset I \subset \mathbb{Z}_2[x_1, \ldots, x_n]$. By Corollary 2.1.7 there exists a Boolean polynomial p s. th. $\mathrm{V}(\langle p, x_1^2 + x_1, \ldots, x_n^2 + x_n \rangle) = \mathrm{V}(I)$. By theorem 2.2.5 we get $I = I(\mathrm{V}(\langle p, x_1^2 + x_1, \ldots, x_n^2 + x_n \rangle)) = \langle p, x_1^2 + x_1, \ldots, x_n^2 + x_n \rangle$. Suppose, there exists a second Boolean polynomial q with $I = \langle q, x_1^2 + x_1, \ldots, x_n^2 + x_n \rangle$. Then

$$V(p) = V(I) = V(q).$$

So p and q define the same characteristic function which means that they are identical Boolean polynomials (theorem 2.1.5).

□

Hence, using theorem 2.1.5, Corollary 2.1.7 and Corollary 2.2.5, we have the following bijections:

$$\begin{gathered} \mathbb{B} \leftrightarrow \{\text{Boolean functions}\} \leftrightarrow \\ \{\text{ideals } I \subseteq \mathbb{Z}_2[x_1, \ldots, x_n] \text{ with } \{x_1^2 + x_1, \ldots, x_n^2 + x_n\} \subseteq I\} \leftrightarrow \\ \{\text{algebraic subsets of } \mathbb{Z}_2^n\} \leftrightarrow \{\text{subsets of } \mathbb{Z}_2^n\}. \end{gathered}$$

Definition 2.2.8 (Boolean Gröbner basis). *For any subset* $H \subseteq \mathbb{Z}_2[x_1, \ldots, x_n]$ *call*

$$\boldsymbol{BI(H)} := \langle H, x_1^2 + x_1, \ldots, x_n^2 + x_n \rangle \subseteq \mathbb{Z}_2[x_1, \ldots, x_n]$$

the ***Boolean ideal of H****. We call a finite set of Boolean polynomials G a* ***Boolean Gröbner basis*** *of H if* $G \cup \{x_1^2 + x_1, \ldots, x_n^2 + x_n\}$ *is a Gröbner basis of* $\mathrm{BI}(H)$*. We call a Boolean Gröbner basis G of H the* ***reduced Boolean Gröbner basis*** *of H, short* **BGB(*H*)** *if there exists* $S \subseteq \{x_1^2 + x_1, \ldots, x_n^2 + x_n\}$*, s. t.* $G \cup S$ *is a reduced Gröbner basis of* $\mathrm{BI}(H)$*. We call a set of Boolean polynomials H a Boolean Gröbner if H is a Boolean Gröbner basis of* $\mathrm{BI}(H)$.

Recall from theorem 2.2.1 that $\mathrm{BGB}(H)$ consists of Boolean polynomials and can be extended to a reduced Gröbner basis of $\mathrm{BI}(H)$ by adding some field polynomials.

Theorem 2.2.9. *Let $p, q \in \mathbb{B}$ with $\mathrm{V}(p) \subset \mathrm{V}(q)$. Then $\langle p, x_1^2 + x_1, \ldots, x_n^2 + x_n \rangle \supset \langle q, x_1^2 + x_1, \ldots, x_n^2 + x_n \rangle$ and we say p implies q. This implication relation forms a partial order on the set of Boolean polynomials.*

Proof. Since both ideals are radical, Hilbert's Nullstellensatz gives the ideal containment. The implication is a partial order by the one-to-one correspondence between Boolean polynomials and sets. It corresponds itself to the inclusion of sets. □

2.2.1 Criteria

Criteria for keeping the set of critical pairs in the Buchberger algorithm small are a central part of any Gröbner basis algorithm aiming at practical efficiency. In most implementations the chain criterion and the product criterion or variants of them are used.

These criteria are of quite general type, and it is a natural question whether we can formulate new criteria for Boolean Gröbner bases. Indeed, this is the case. There are two types of pairs to consider: Boolean polynomials with field equations, and pairs of Boolean polynomials. We concentrate on the first kind of pairs here.

Theorem 2.2.10 (Linear lead factor criterion). *Let $f \in \mathbb{B}$ be of the form $f = l \cdot g$, l a polynomial with linear leading term x_i, and $g \in \mathbb{Z}_2[x_1, \ldots, x_n]$ be any polynomial. Then* $\mathrm{spoly}(f, x_i^2 + x_i)$ *has a nontrivial t-representation (definition 1.2.3) against $\{f, x_1^2 + x_1, \ldots, x_n^2 + x_n\}$.*

Proof. First, we consider the case $g = 1$. In this situation the following formula holds: $\mathrm{lm}(f) = x_i$. Let r be a reduced normal form of $\mathrm{spoly}(f, x_i^2 + x_i)$ against f and the field equations. Then r is (tail) reduced, so it is a Boolean polynomial and irreducible against f, so x_i does not occur. In particular considered as a Boolean function it is independent from the value of x_i.

Since r is a linear combination of f and field equations (which are zero considered as Boolean functions) we get:

$$r(x_1, \ldots, x_n) = 1 \Rightarrow f(x_1, \ldots, x_n) = 1.$$

Now, we assume that $r \neq 0$. As a non-zero Boolean polynomial corresponds to a non-zero Boolean function, we know that there exist $v_1, \ldots, v_n \in \{0, 1\}$ subject to $r(v_1, \ldots, v_n) = 1$. The above implication gives $f(v_1, ..., v_n) = 1$.

Then we can change the value of x_i without affecting the value of r

$$r(v_1, \ldots, v_i + 1, \ldots, v_n) = 1,$$

but

$$f(v_1, \ldots, v_i + 1, \ldots, v_n) = 0$$

as x_i only occurs in the one term x_i of f. This contradicts the above implication between r and f. So $r = 0$ and $\mathrm{spoly}(f, x_i^2 + x_i)$ has a standard representation.

Now, we consider a general Boolean polynomial $g \neq 1$ ($f = l \cdot g$, $\deg(\mathrm{lm}(l)) = 1$): $\mathrm{spoly}(l, x_i^2 + x_i)$ has a standard representation against l and the field equations:

$$\mathrm{spoly}(l, x_i^2 + x_i) = \sum_{j=1}^{n} h_j \cdot x_j^2 + x_j + \alpha \cdot l\,,$$

for polynomials α, h_j $(j \in \{1, \ldots, n\})$:

$$x_j^2 \cdot \mathrm{lm}(h_j) \le \mathrm{lm}(\mathrm{spoly}(l, x_i^2 + x_i)) < x_i^2, \quad \mathrm{lm}(\alpha \cdot x_i) < x_i^2.$$

We multiply this equation by g and get by that fact that x_i does not occur in g:

$$\begin{aligned}\mathrm{spoly}(l \cdot g, x_i^2 + x_i) &= \mathrm{spoly}(l \cdot g, g \cdot x_i^2 + x_i) - \mathrm{tail}(g) \cdot (x_i^2 + x_i) \\ &= g \cdot \mathrm{spoly}(l, x_i^2 + x_i) - \mathrm{tail}(g) \cdot (x_i^2 + x_i).\end{aligned}$$

Using the standard representation for $\mathrm{spoly}(l, x_i^2 + x_i)$ from above, both summands have a t-representation for a monomial $t < x_i^2 \cdot \mathrm{lm}(g)$, so we also get a nontrivial t-representation in the sum. □

Lemma 2.2.11. *Let G be a Gröbner basis, f a polynomial, then $\{f \cdot g | g \in G\}$ is Gröbner basis.*

Remark 2.2.12. *This lemma is trivial, we just want to show the difference to the next theorem.*

Theorem 2.2.13. *Let G be a reduced Boolean Gröbner basis, $l \in \mathbb{B}$ with $\deg(\mathrm{lm}(l)) = 1$ and $\mathrm{vars}(l) \cap \mathrm{vars}(g) = \emptyset$ for all $g \in G$. Then $\{l \cdot g | g \in G\}$ is a reduced Boolean Gröbner basis,that is, $\{l \cdot g | g \in G\} \cup \{x_1^2 + x_1, \ldots, x_n^2 + x_n\}$ is a Gröbner basis. In other words, we get a Gröbner basis by multiplying the Boolean polynomials, but not the field equations with the special polynomial l. Note, that $\mathrm{BI}(\{l \cdot g | g \in G\})$ is in general larger than the ideal $\{l \cdot p | p \in \mathrm{BI}(G)\}$.*

Proof. We show that every S-Polynomial has a non-trivial t-representation (theorem 1.3.3). We have to consider three types of pairs. If p, q are both field polynomials, $\mathrm{spoly}(p, q)$ has a standard representation by the product criterion. If p, q are both Boolean polynomials, then $\mathrm{spoly}(l \cdot p, l \cdot q)$ has a standard representation by multiplying the standard representation of $\mathrm{spoly}(p, q)$ by l. Now let p be a Boolean polynomial and q a field polynomial, say $q = x^2 + x$. If $\mathrm{lm}(l) = x$, then $\mathrm{spoly}(l \cdot p, q)$ has a nontrivial t-representation by theorem 2.2.10. If x occurs in $\mathrm{lm}(p)$, then by lemma 2.2.11 $\mathrm{spoly}(l \cdot p, l \cdot q)$ has a standard representation against $\{l \cdot g | g \in G\} \cup \{l \cdot e | e \in \{x_1^2 + x_1, \ldots, x_n^2 + x_n\}\}$, so also against the set $\{l \cdot g | g \in G\} \cup \{x_1^2 + x_1, \ldots, x_n^2 + x_n\}$. Hence, we just have to show that the difference to $\mathrm{spoly}(l \cdot p, l \cdot q)$ has a t-representation with $t < \mathrm{lm}(p) \cdot \mathrm{lm}(l) \cdot x := c$. Setting

$$h := \mathrm{spoly}(l \cdot p, l \cdot (x^2 + x)) - \mathrm{spoly}(l \cdot p, x^2 + x) = \mathrm{tail}(l) \cdot (x^2 + x)$$

we get that $x^2 + x$ divides h, and $\mathrm{lm}(h) = \mathrm{lm}((x+1) \cdot \mathrm{tail}(l)) \cdot x < c$ since $\mathrm{lm}(p)$ contains x. So h has standard representation against $x^2 + x$. If x does neither occur in $\mathrm{lm}(f)$ nor in $\mathrm{lm}(l)$ the product criterion applies. Reducedness follows from the fact that l does not share any variables with G. □

2.2.2 Symmetry and Boolean Gröbner bases

In this section we will show how to use the theory presented in the previous section to build faster algorithms by using symmetry and simplification by pulling out factors with linear leads.

For a polynomial p we denote by $\mathrm{vars}(p)$ the set of variables actually occurring in the polynomial.

Definition 2.2.14. *Let p be a polynomial in $\mathbb{Z}_2[x_1,\ldots,x_n]$ with a given monomial ordering $>$, $|\operatorname{vars}(p)| = k$, $I = \operatorname{vars}(p) = \{x_{i_1},\ldots,x_{i_k}\}$, and $J = \{x_{j_1},\ldots,x_{j_k}\}$ be any set of k variables. We call a morphism of polynomials algebras over $\mathbb{Z}_2$*

$$f : \mathbb{Z}_2[I] \to \mathbb{Z}_2[J] : x_{i_s} \mapsto x_{j_s} \text{ for all } s\,,$$

a suitable shift for the variables of p if and only if for all monomials $t_1, t_2 \in \mathbb{Z}_2[I]$ the relation $t_1 > t_2 \iff f(t_1) > f(t_2)$ holds.

Remark 2.2.15. *In the following we concentrate on the problem of calculating $\operatorname{BGB}(p)$ for one Boolean polynomial p (non-trivial as field equations are implicitly included). So, if we know $\operatorname{BGB}(q)$ for a Boolean polynomial q and if there exists a suitable shift f with $f(q) = p$, then $f(\operatorname{BGB}(q)) = \operatorname{BGB}(p)$. Hence, we can avoid the computation of $\operatorname{BGB}(p)$. Adding all elements of $BGB(p)$ to our system means that we can omit all pairs of the form $(p, x_i^2 + x_i)$. A special treatment (using caching and tables) of this kind of pairs is a good idea because this is an often reoccurring phenomenon. As these pairs depend only on p (the field equations are always the same), this reduces the number of pairs to consider significantly.*

Remark 2.2.16. *Note that the concept of Boolean Gröbner bases fits very well here as $\operatorname{BGB}(p)$ is the same in $\mathbb{Z}_2[\operatorname{vars}(p)]$ as in $\mathbb{Z}_2[x_1,\ldots,x_n]$ although the last case refers to a Gröbner basis with more field equations.*

Definition 2.2.17. *We define the relation $p \sim_{pre} q$ if and only if there exists a suitable shift between p and q or if there exists an l with $\deg(\operatorname{lm}(l)) = 1$ and $p = l \cdot q$. From $\sim_{pre}$ we derive the relation $\sim_{sym}$ as its reflexive, symmetric, transitive closure (the smallest equivalence relation containing $\sim_{pre}$).*

Remark 2.2.18. *For all p and q in an equivalence class of $\sim_{sym}$ the Boolean Gröbner basis $\operatorname{BGB}(p)$ can be mapped to $\operatorname{BGB}(q)$ by a suitable variable shift and pulling out (or multiplying) by Boolean polynomials with linear lead. In practice, we can avoid complete factorizations by restricting ourselves to detect factors of the form x or $x+1$. Using these techniques it is possible to avoid the explicit calculation of many critical pairs.*

Definition 2.2.19. *A monomial ordering is called* ***symmetric*** *if the following holds. For every k, and every two subsets of variables $I = \{x_{i_1},\ldots,x_{i_k}\}$, and $J = \{x_{j_1},\ldots,x_{j_k}\}$ with $i_z < i_{z+1}$, $j_z < j_{z+1}$ for all $z = 1,\ldots,k$ the $\mathbb{Z}_2$-algebra homomorphism*

$$f : \mathbb{Z}_2[I] \to \mathbb{Z}_2[J] : x_{i_z} \mapsto x_{j_z}$$

defines a suitable shift.

For a symmetric ordering it is always possible to map a polynomial p to the variables $x_1,\ldots,x_{|\operatorname{vars}(p)|}$ by a suitable shift. This is utilized in algorithm 5 for speeding up calculation of Boolean Gröbner bases. In the following we assume that the representative chosen in the algorithm is canonical (in particular uniquely determined in the equivalence class in $\sim_{sym}$), if every factor with linear lead is pulled out.

Algorithm 5 Calculating BGB(p) in a symmetric order

Input: $p \in \mathbb{B}$, $>$ a monomial ordering

Output: BGB(p)

pull out as many factors with linear lead as possible
calculate a more canonical representative q of the equivalence class of p in $\sim_{sym}$ by shifting p to the first variables
if q lies in a cache or table **then**
  $B :=$ BGB(q) from cache
else
  $B :=$ BGB(q) by Buchberger's algorithm
shift B back to the variables of p
multiply B by the originally pulled out factors
return B

Remark 2.2.20. *From the implementation point of view, it turned out to be useful to store the* BGB *of all* 2^{16} *Boolean polynomials in up to four variables in a precomputed table, for more variables we use a dynamic cache (pulling out factors reduces the number of variables). Using canonical representatives increases the number of cache hits.*

The technique for avoiding explicit calculations can be integrated in nearly every algorithm similar to the Buchberger's algorithm. Best results were made by combining these techniques with the algorithm slimgb [Brickenstein, 2010]. We call this combination symmgbGF2. *For our computations the strategy in slimgb for dealing with elimination orderings is quite essential.*

Practical meaning of symmetry techniques

The real importance of symmetry techniques should not only be seen in avoiding computations in leaving out some pairs. In contrast, application of the techniques described above changes the behaviour of the algorithm completely. Having a Boolean polynomial p, the sugar value [Giovini et al., 1991] of the pair $(p, x^2 + x)$ is usually $\deg(p) + 1$ which corresponds to the position in the waiting queue of critical pairs. It often occurs that in BGB(p) polynomials with much smaller degree occur.

Having these polynomials earlier, we can avoid many other pairs in higher degree. This applies quite frequently in this area, in particular when we have many variables, but the resulting Gröbner basis looks quite simple (for example linear polynomials). The earlier we have these low degree polynomials, the easier the remaining computations are, resulting in less pairs and faster normal form computations.

2.2.3 Greedy normal form algorithm

In this section, we will introduce the greedy normal form (algorithm 6) which is a variant of the Buchberger normal form (algorithm 1). This normal form algorithm is particularly suited for the case of Boolean polynomials as we will show in theorem 2.2.21.

The relevance of the greedy normal form algorithm is underlined by the algorithm llnf* (section 4.3) which can be seen as a specialization of the greedy normal form algorithm: It handles a special family of polynomial systems and has applications in formal verification.

Algorithm 6 Greedy normal form

Input: G finite tuple of Boolean polynomials, f Boolean polynomial.
Output: $\mathrm{REDNF}(f, G \cup \{x_1^2 + x_1, \ldots, x_n^2 + x_n\})$
  while $f \neq 0$ and $(\exists g \in G : \mathrm{lm}(g) | \mathrm{lm}(f))$ **do**
    $h := \frac{f}{\mathrm{lm}(g)}$ /* division by remainder definition 2.1.16 */
      /* s. th. the result corresponds to terms in f divisible by $\mathrm{lm}(g)$ */
    $f := f - h \star g$ /* no term of f is divisible by $\mathrm{lm}(g)$ any more */
  return f

Algorithm 6 combines many small steps into one. The cost of the single steps may be higher using ZDD operations (chapter 3), but the combined step can be done much faster. The high cost (compared to classical polynomial representations) of these single additions might be surprising in the first, but can be explained quite easily. Good normal form strategies try to select a monomial for g whenever possible. Then for classical structures like linked lists one does not need to use a general addition routine, but can simply pop the first element (term) from the list. This can be done in constant time. In fact only applying this greedy technique to the case, where g is a monomial, already gives a quite good normal form implementation in POLYBORI. But also for classical data structures we see advantages of this normal form algorithm in case of Boolean polynomials as for the multiplication step asymptotical fast variants can be used. Of course, it is a matter of heuristics to decide when it might be better to perform a single reduction step.

Theorem 2.2.21. *Let f, g be Boolean polynomials in $\mathbb{Z}_2[x_1, \ldots, x_n]$, $\mathrm{lm}(g) | \mathrm{lm}(f)$. Then*

$$r := f - \frac{f}{\mathrm{lm}(g)} \cdot g$$

does not contain any term which is a multiple of $\mathrm{lm}(g)$: $\frac{r}{\mathrm{lm}(g)} = 0$. In particular, this claim holds for

$$r_2 := f - \frac{f}{\mathrm{lm}(g)} \star g$$

as well. Additionally, $\mathrm{supp}(r) \not\supset \mathrm{supp}(\mathrm{spoly}(f, g))$ if $r \neq \mathrm{spoly}(f, g)$.

Proof. We only have to show that $\frac{f}{\mathrm{lm}(g)} \cdot \mathrm{tail}(g)$ does not contain multiples of $\mathrm{lm}(g)$ as $\frac{f}{\mathrm{lm}(g)} \cdot \mathrm{lm}(g)$ (polynomial consisting of the terms of f divisible by $\mathrm{lm}(g)$) cancels with all such terms in f. We may assume $\mathrm{lm}(g) \neq 1$ (in this case $\mathrm{tail}(g) = 0$). Then by elementary properties of monomial orderings no term in $\mathrm{tail}(g)$ is divisible by $\mathrm{lm}(g)$. On the other hand every term in $\frac{f}{\mathrm{lm}(g)}$ does not contain a variable occurring in $\mathrm{lm}(g)$ (the maximal exponent per variable is 1), so multiplying with this terms does not contribute to divisibility by $\mathrm{lm}(g)$. This proves the first claim. The last claim can be shown analogously. □

The algorithm only makes sense for Boolean polynomials. In fact, in the non-Boolean case a single reduction step can give worse results and would be harder from the computational point of view than a reduction step in the classical Buchberger normal form algorithm.

Example 2.2.22. *We consider the polynomials $f = x^2 + x \cdot y$ and $g = x + y$ in $\mathbb{Q}[x, y]$ with lexicographical ordering ($x > y$). Then* $\operatorname{spoly}(f, g) = 0$, *but*

$$f + \frac{f}{\operatorname{lm}(g)} \cdot g = f - (x + y) \cdot g = -x \cdot y - y^2.$$

The purpose of the the construction was to cancel all multiples of $\operatorname{lm}(g)$ *in f which includes $x \cdot y$ in this example. Apparently the strategy fails in this general case.*

2.3 Intersection of ideals in Boolean rings

Theorem 2.3.1. *Given two ideals $I_1, I_2 \subset K[x_1, \ldots, x_n]$, both containing $\langle x_1^2 + x_1, \ldots, x_n^2 + x_n\rangle$ with generators $F = \{f_i | i \in \{1, \ldots, r\}\}$, respectively $G = \{g_i | i \in \{1, \ldots, s\}\}$. Consider the ideal $J = \langle t \cdot (t-1), t \cdot f_1, \ldots, t \cdot f_r, (t-1) \cdot g_1, \ldots, (t-1)t \cdot g_s\rangle \subset K[x_1, \ldots, x_n, t]$. Then $I_1 \cap I_2 = J \cap K[x_1, \ldots, x_n]$.*

Proof. Exactly as in Lemma 1.8.10 in [Greuel and Pfister, 2002]. □

Remark 2.3.2. *The previous theorem differs from the classical trick just by adding the equation $t^2 + t$. While this seems to be in the general case only an optimization, this is strictly necessary if we want to calculate the intersection in a Boolean ring (with field equations for every variable implicitly contained in every ideal). The other field equations of the form $x_i^2 + x_i$ are contained in both ideals anyway and we have*

$$t \cdot (x_i^2 + x_i) + (t - 1) \cdot (x_i^2 + x_i) = x_i^2 + x_i.$$

Algorithm 7 Intersection of Boolean ideals

Input: $I_1 = \langle f_1, \ldots, f_r, x_1^2 + x_1, \ldots, x_n^2 + x_n\rangle$, $I_2 = \langle g_1, \ldots, g_r, x_1^2 + x_1, \ldots, x_n^2 + x_n\rangle$, f_i, g_j Boolean polynomials
Output: result Boolean generators $I_1 \cap I_2$ modulo field equations
add a variable t
form an elimination order for t
$R := \operatorname{BGB}(t \cdot f_1, \ldots, t \cdot f_r, (t-1) \cdot g_1, \ldots, (t-1)t \cdot g_s)$
return all elements in R which do not involve the variable t

Remark 2.3.3. *The nice thing about this algorithm is that it just reads like the classical versions using the concept of a Boolean Gröbner basis. However, there exist modifications in the treatment of field equations which must not be forgotten.*

Chapter 3

Polynomial representations as ZDD

In this chapter we introduce a representation of Boolean polynomials as zero-suppressed decision diagrams (ZDDs). It starts with some basic definitions and properties of directed graphs and decision diagrams. After that zero-suppressed decision diagrams are introduced and it is described how to interpret them as polynomials. Moreover, it is shown how monomial orderings correspond to the ZDD structure.

3.1 Formal introduction to decision diagrams

In the following, we will define the basic concepts of decision diagrams. We start with some basic concepts from graph theory.

3.1.1 Directed graphs

For introducing directed graphs, we follow the notations in [Clark and Holton, 1991] as close as possible.

Definition 3.1.1. *A* ***directed graph*** *(digraph) is a tuple $D = (N, A)$ consisting of a nonempty set of nodes N and a set of arcs A where every arc $a \in A$ is associated with a pair (u, v) of nodes.*

- *u is called the* ***origin*** *of a: $u = \mathrm{origin}(a)$*
- *v is called the* ***terminus*** *of a: $v = \mathrm{terminus}(a)$*

Definition 3.1.2. *Let $D = (N, A)$ be a digraph, k a positive integer, $a_1, \ldots a_n \in A$,*

$$\mathrm{origin}(a_i) = \mathrm{terminus}(a_{i-1}) \textit{ for all } i > 2, i \leq k.$$

Then

$$\mathrm{origin}(a_1), a_1, \mathrm{terminus}(a_1), a_2, \mathrm{terminus}(a_2), \ldots a_k, \mathrm{terminus}(a_k)$$

is called a ***(directed) walk*** *in D. Furthermore,* $\mathrm{origin}(a_1)$ *is called* ***start point*** *and* $\mathrm{terminus}(a_k)$ ***end point*** *of the walk.*

Definition 3.1.3. *A digraph D is called* ***acyclic*** *if there exists no directed walk in it where the start point is equal to its end point.*

A node v in a digraph is said to be ***reachable*** *from a node u if there exists a directed walk with start point u and end point v.*

The digraph is called ***rooted*** *if there exists exactly one node that is not the terminus of an arc and all other nodes are reachable from that node. This node is called the* ***root*** *of D: $\text{root}(D)$.*

A ***directed path*** *is a directed walk where all vertices are distinct.*

Remark 3.1.4. *In an acylic digraph each directed walk is a directed path.*

Definition 3.1.5. *A* ***morphism*** *γ between directed graphs $D_1 = (N_1, A_1)$ and $D_2 = (N_2, A_2)$ consists of two maps $\alpha : N_1 \longrightarrow N_2, \beta : A_1 \longrightarrow A_2$ ($\gamma = (\alpha, \beta)$), where*

$$\alpha(\text{origin}(a)) = \text{origin}(\beta(a)) \text{ and } \alpha(\text{terminus}(a)) = \text{terminus}(\beta(a))$$

for all $a \in A_1$. It is called an ***isomorphism*** *if the maps are bijective.*

Definition 3.1.6. *Let $D = (N, A)$ be a directed graph. Then a directed graph $E = (O, B)$ is a* ***(directed) subgraph*** *of D if $O \subseteq N$ and $B \subseteq A$.*

3.1.2 Binary Decision Diagrams

Definition 3.1.7 (Binary Decision Diagram, BDD). *An* ***ordered binary decision diagram*** *(BDD) is a rooted, directed, and acyclic graph $D = (N, A)$ with additional structure. Formally it can be defined as a 7-tuple:*

- *finite set $N_N \subset N$ (**non-terminal nodes**),*
- *finite set $N_T \subseteq \{0, 1\} \subseteq N$ (**terminal nodes**),*
- *finite set $A_T \subseteq A$ (**then-arcs/edges**),*
- *finite set $A_E \subseteq A$ (**else-arcs/edges**),*
- *finite set V (**variables**),*
- $\text{var} : N_N \longrightarrow V$ *(non-terminal nodes correspond to variables) and*
- $\text{vi} : V \hookrightarrow \mathbb{N}$*, injective (each variable has an index).*

We demand the following properties:

- *$N = N_T \cup N_N$, $N_T \cap N_N = \emptyset$: a node is either "terminal" or "non-terminal".*
- *$A = A_E \cup A_T$, $A_E \cap A_T = \emptyset$: an arc is either an "else" or a "then" arc.*
- *For all $n \in N_N$: exists exactly one $e \in A_E, t \in A_T$: $\text{origin}(t) = n, \text{origin}(e) = n$ we say: $e = \text{elsearc}(n)$, $t = \text{thenarc}(n)$, there are two arcs outgoing from every non-terminal node: an else and an then-arc.*

- *For all* $t \in N_T \nexists a \in A_E \cup A_T : \text{origin}(a) = t$*: Terminal nodes really terminate paths: There is no arc outgoing from them.*

- *For all* $a \in A_E \cup A_T$ *: if* $\text{terminus}(a) \in N_N$ *then* $\text{vi}(\text{var}(\text{origin}(a))) < \text{vi}(\text{var}(\text{terminus}(a)))$*: Following a directed walk, the variables appear ascending by their index.*

- $\text{vi}(V) = \{1, 2, \ldots, |V|\}$

- $\exists_1 n \in N_N \cup N_T : \nexists a \in A_E \cup A_T : n = \text{terminus}(a)$*: the graph has a unique root.*

The corresponding digraph is derived from this tuple as

$$(\overbrace{N_N \cup N_T}^{=N}, \overbrace{A_T \cup A_E}^{=A}).$$

with two terminal nodes $\{0, 1\}$ *and decision nodes.*

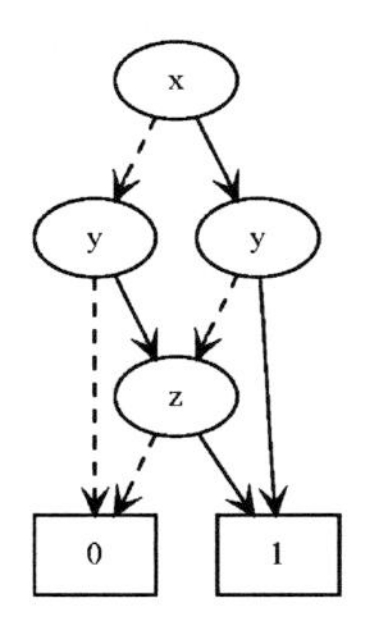

Figure 3.1: Example diagram

Remark 3.1.8. *While defining a 7-tuple does not look very pleasant, all components are immediately recognizable in the simple example in figure 3.1.*

- *The non-terminal nodes* N_N *are the nodes with round shape while the terminal node* N_T *have rectangular shape.*

- *The then-arcs* A_T *are drawn solid while the else-arcs* A_E *are visualized by a dashed line.*

- *The variables usually correspond to the variables in the polynomial ring and the mapping of non-terminal nodes to variables is visualized by writing the variable as label in the node.*

- *Finally the indexing of variables is implicitly given by the order of the variables appearing in the diagram (from top to bottom).*

Typically V will be the set of variables in the polynomial ring.

Definition 3.1.9. *$\gamma = (\alpha, \beta)$ defines an isomorphism between ordered binary decision diagrams $B_1 = (N_{N_1}, N_{T_1}, A_{T_1}, A_{E_1}, V_1, \mathrm{var}_1, \mathrm{vi}_1)$ and $B_2 = (N_{N_2}, N_{T_2}, A_{T_2}, A_{E_2}, V_2, \mathrm{var}_2, \mathrm{vi}_2)$ if the following holds:*

- *γ is an isomorphism of directed graphs,*
- $V_1 = V_2$,
- $\mathrm{vi}_1 = \mathrm{vi}_2$,
- $\alpha(N_{N_1}) = N_{N_2}$,
- $N_{T_1} = N_{T_2}$,
- *$\alpha(t) = t$ for all $t \in N_{T_1}$,*
- $\beta(A_{T_1}) = A_{T_2}$,
- *$\beta(A_{E_1}) = A_{E_2}$ and*
- *$\mathrm{var}_1(n) = \mathrm{var}_2(\alpha(n))$ for all $n \in N_{N_1}$.*

Definition 3.1.10. *Let $D = (N_N, N_T, A_T, A_E, V, \mathrm{var}, \mathrm{vi})$ be a BDD, $n \in N_N \cup N_T$. Let N be the set of nodes of D reachable from n including n itself. Define*

$$A := \{a \in A_E \cup A_T | \operatorname{terminus}(a) \in N \text{ and } \operatorname{origin}(a) \in N\}.$$

Then

$$\operatorname{subdiag}(n) := (N_N \cap N, N_T \cap N, A_T \cap A, A_E \cap A, V, \mathrm{var}_{|N_N \cap N}, \mathrm{vi})$$

*is called the **subdiagram of D outgoing from n**.*

Definition 3.1.11. *An ordered binary decision diagram is called **reduced** if it contains no distinct subgraphs that are isomorphic as ordered binary decision diagrams.*

Example 3.1.12. *Definition 3.1.11 is illustrated by figure 3.2(a): The diagrams outgoing from the z-nodes in this figures are isomorphic. The correct reduced version of this graph is the one from the introductory example in Figure 3.1.*

Definition 3.1.13. *A reduced ordered decision diagram $D = (N_N, N_T, A_T, A_E, V, \mathrm{var}, \mathrm{vi})$ is called **zero-suppressed decision diagram (ZDD)** if $\operatorname{terminus}(a) \neq 0$ for all then arcs $a \in A_T$.*

This implies, that all then arcs leading to the zero terminal node are left out implicitly.

Example 3.1.14. *Definition 3.1.13 is illustrated in figure 3.3. The introductory example in figure 3.1 was extended by an artificial z-node which is forbidden for ZDDs.*

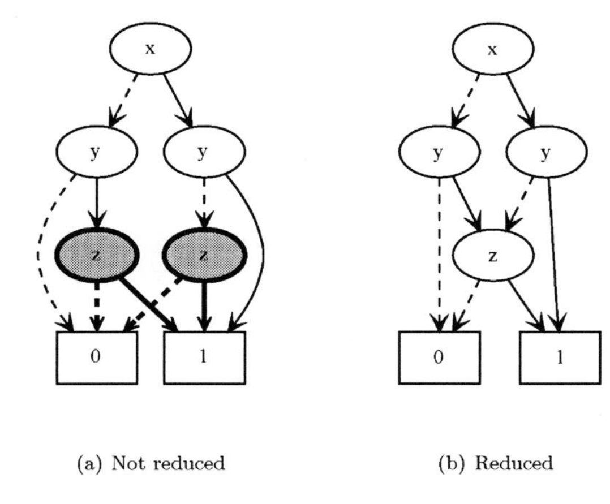

(a) Not reduced (b) Reduced

Figure 3.2: Reduced and not reduced diagrams

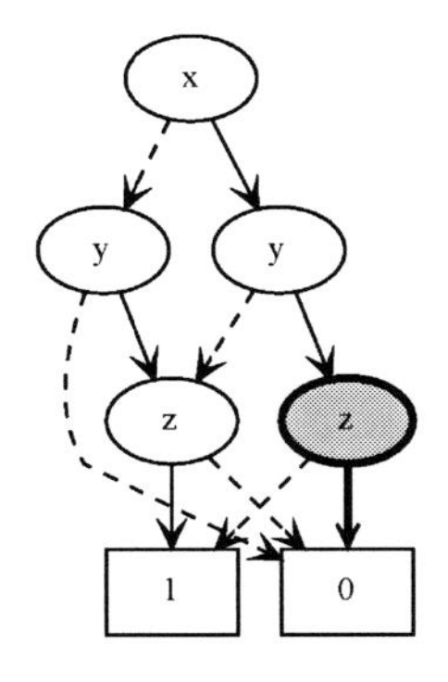

Figure 3.3: No then-arc is allowed to point to 0

Definition 3.1.15. *Let b be a binary decision diagram.*

- *The* ***decision variable*** *associated to the root node of b is denoted by*

$$\textbf{topvar}(\boldsymbol{b}) := \mathrm{var}(\mathrm{root}(b)).$$

*The **index of the root node** is denoted by*

$$\mathbf{top}(\boldsymbol{b}) := \mathrm{vi}(\mathrm{topvar}(b)).$$

If the root node is a terminal node (0 or 1), we define $\mathrm{top}(b) = |V|+1$. *Furthermore,* b^T *(**then branch**) and* b^E *(**else branch**) indicate the (sub-)diagrams outgoing from the terminus of the then- and else-arc, respectively, of the root node of b (see figure 3.4)*

- *For two BDDs* b_1, b_0 *which do not depend on the decision variable* x *the **if-then-else operator*** $\mathrm{ite}(x, b_1, b_0)$ *denotes the BDD c which is obtained by introducing a new node associated to the variable* x, *s. th.* $c^T = b_1$, *and* $c^E = b_0$. *Note, that the index of* x *has to be less than* $\mathrm{top}(b_0)$, $\mathrm{top}(b_1)$.

Remark 3.1.16. *Algorithms operating directly on binary decision diagrams are usually recursive. Typically in each call* top *increases. Since it can only take* $|V| + 1$ *different values, the recursion depth is bounded. So algorithms will always terminate (when using only recursion following this principle, conditions and terminating function call). Moreover, we can always prove correctness by induction on the minimal* top *of all function arguments.*

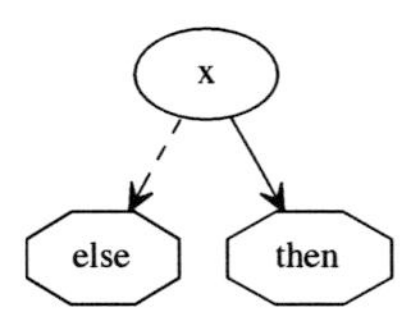

Figure 3.4: Branches of a diagram

Definition 3.1.17. *Let* $b = (N_N, N_T, A_T, A_E, V, \mathrm{var}, \mathrm{vi})$ *be a ZDD and let*

$$p = (n_1, a_1, n_2, a_2 \ldots, n_k)$$

be a directed path in b starting at the root node of b

$$n_1 = \mathrm{root}(b)$$

ending with

$$n_k \in \{0, 1\}.$$

*Then the sequence p is called a **terminated path** of b. If* $n_k = 1$, *it is called a **valid path**. For a valid path p we define*

- *the* ***set of*** p

$$\mathbf{set}(\boldsymbol{p}) = \{\text{var}(n_i) | a_i \in A_T\},$$

- *the* ***monomial or term of*** p

$$\mathbf{term}(\boldsymbol{p}) = \prod_{v \in \mathsf{set}(p)} v$$

(if V is part of a multiplicative monoid).

Definition 3.1.18. *Given $b = (N_N, N_T, A_T, A_E, V, \text{var}, \text{vi})$ we denote the set of all valid paths in b by* paths(b).

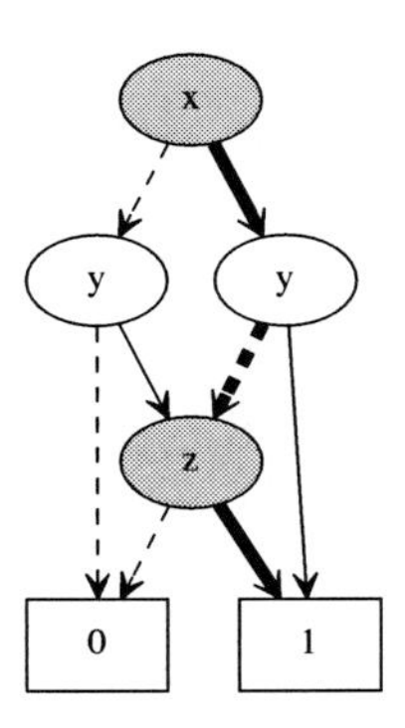

Figure 3.5: The monomial $x \cdot z$ is associated with the highlighted path

Remark 3.1.19. *Definition 3.1.17 means that the set/monomial associated to a valid path only contains variables of nodes where the path follows the outgoing* **then edge**. *This is illustrated in figure 3.5.*

Definition 3.1.20. *Using definition 3.1.17 we can make use of a ZDD $b = (N_N, N_T, A_T, A_E, V, \text{var}, \text{vi})$ to represent various mathematical objects. We will assume that V is a subset of a ring, typically the set of variables in a polynomial ring.*

- *The* ***Boolean Set***

$$\mathbf{booleanset}(\boldsymbol{b}) := \{\text{set}(p) | p \textit{ valid path in } b\},$$

- *the* ***polynomial***

$$\mathbf{polynomial}(\boldsymbol{b}) := \sum_{p \textit{ valid path}} \left(\prod_{v \in \mathsf{set}(p)} v \right)$$

.

- *the* ***support*** *of* polynomial(b), *which can be constructed directly as*

$$\mathbf{supp(polynomial(}\boldsymbol{b}\mathbf{))} := \left\{ \left(\prod_{v \in \mathrm{set}(p)} v \right) | p \textit{ valid path} \right\},$$

- *a* ***set of vectors*** *in* $\{0,1\}^{|V|}$*:*

$$\mathbf{vectorset(}\boldsymbol{b}\mathbf{)} := \{\sum_{x_i \in v} e_i | v \in \mathrm{booleanset}(b)\},$$

  where e_i *denotes the* i*-th unit vector.*

Of course booleanset(b), polynomial(b) *and* supp(b) *can be identified with each other.*

More on the consequences of these graph-theoretical aspects of decision diagrams can be found in [Bryant, 1986].

In the following it is always assumed that the variables of the ZDDs are the variables of a multivariate polynomial ring over $\mathbb{Z}_2$.

Remark 3.1.21. *It is also possible to refer to these diagrams as functional decision diagrams (FDDs, see [Kebschull et al., 1992]). While we store the term structure of a polynomial in a ZDD, it results in the same graph like in the case when one stores the polynomial function in an FDD. From a computational algebra point of view, our description is more natural (and also used in [Minato, 1995, Chai et al., 2008]). From a verification point of view the FDD is the more appropriate description.*

3.2 Introduction to ZDDs

This section will introduce and visualize polynomial representations as ZDD.

We start with a variable x. The corresponding ZDD consists of a single node and is shown in figure 3.6(a)

3.2.1 Variation by multiplication

Multiplying by y we obtain the picture for $x \cdot y$ in Figure 3.6(b). Analogously figure 3.6(c) represents $x \cdot y \cdot z$.

In this way, we see that a monomial looks like a single path through all its variables leading to the 1 terminal node.

3.2.2 Variation by addition

After getting a picture, how multiplication of variables look like, we continue with addition: We choose the polynomial $x + y$.

As we can see in figure 3.10(a), there are now two path leading to the terminal 1-node.

This extends to the representation of $x + y + z$ (figure 3.10(b)) in a natural way. This polynomial has three terms. Hence the ZDD has three paths leading to the terminal 1-node.

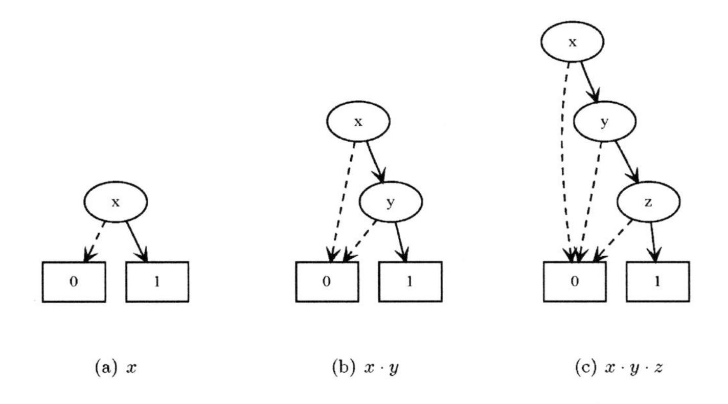

(a) x (b) $x \cdot y$ (c) $x \cdot y \cdot z$

Figure 3.6: Variation by multiplication

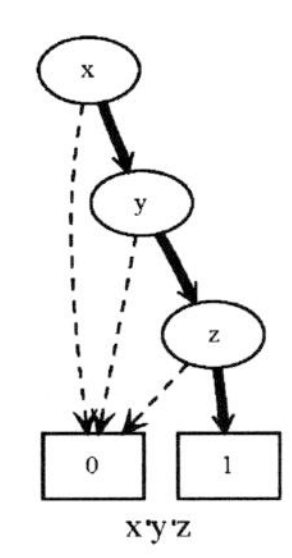

Figure 3.7: **Terms** of $x \cdot y \cdot z$

3.2.3 More polynomials

After getting a fundamental understanding for these very simple polynomials, we try to understand more complicated examples. We add 1 to the above presented polynomial $x \cdot y \cdot z$ and get the picture shown in Figure 3.11 for the terms $x \cdot y \cdot z + 1$:

As we can see, we have instead all else-edges pointing to the terminal 0-node a valid path pointing to terminal 1-node. This can be obtained by choosing the else-branch from the root node which yields in some sense the empty path corresponding to the term 1.

Now we have a look at the changes if we use the polynomial $x \cdot y \cdot z + x$ instead (which means multiplying the summand 1 by x).

As we can see in Figure 3.12, the else-edge leading to the terminal one-node moves

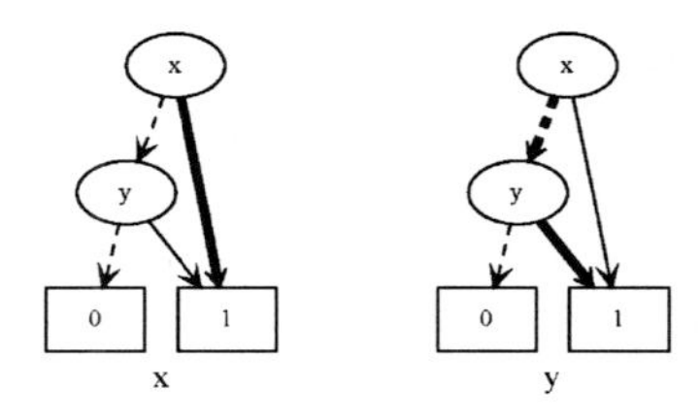

Figure 3.8: **Terms** of $x + y$

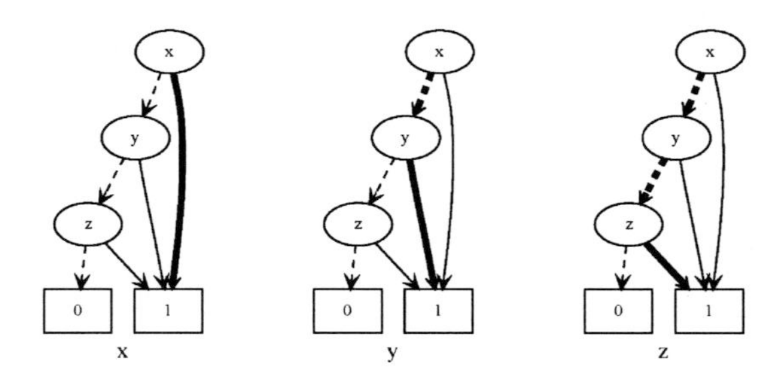

Figure 3.9: **Terms** of $x + y + z$

down as this term now starts with x.

We continue the game with the polynomial $x \cdot y \cdot z + y$ (Figure 3.13)

In the first moment, the graph might look surprisingly different. In reality the path describing the second summand, just changes the order of its branches: First we take the else-branch at the node representing the variable x (as the term y does not contain x), then the then-branch at y. In Figure 3.12 we had just the opposite order of branches.

Finally, we take a look at $x \cdot y \cdot z + z$ in the hope to detect something familiar (Figure 3.14).

In comparison to the previous example, the left y-node is replaced by a z-node which is merged with the z-node in the right part of Figure 3.13.

3.2.4 Special sets

In this section a few examples of very structured polynomials are presented which feature also very structured diagrams. The first example chosen in this section is the polynomial $(a + 1) \cdot (b + 1) \cdot (c + 1) \cdot (d + 1) \cdot (e + 1)$. It contains every monomial in the variables a,

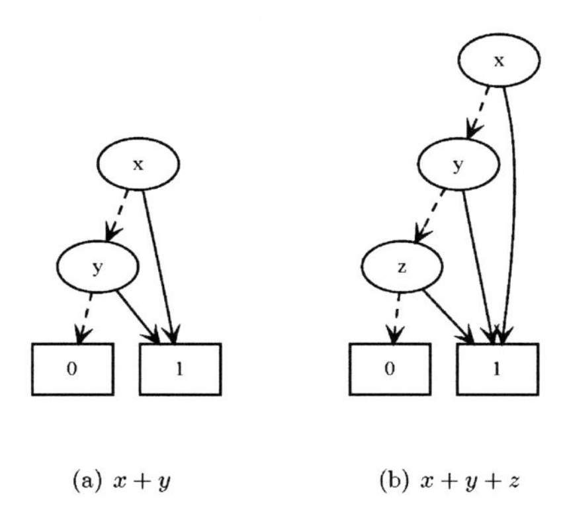

(a) $x + y$ (b) $x + y + z$

Figure 3.10: Variation by addition

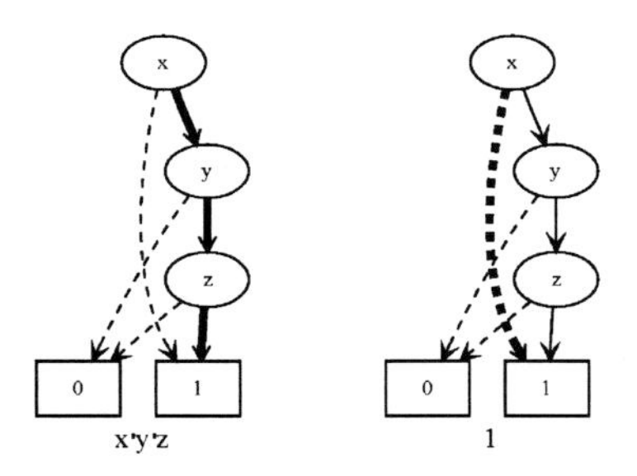

Figure 3.11: **Terms** of $x \cdot y \cdot z + 1$

b, c, d, e: Figure 3.15.

The polynomial has 32 terms, but the graph contains only 5 non-terminal nodes and 10 edges. Starting from the root node, on each of the five levels we can choose the then- or else-branch and will always end at the terminal 1-node (so get a valid path). This means that we have five times two possibilities, so we get 32 combinations which is exactly the number of terms.

In the next example, we restrict the degree for each term to a maximum of two to represent the polynomial: $a \cdot (b+c+d+e)+b \cdot (c+d+e)+c \cdot (d+e)+d \cdot e+a+b+c+d+e+1$ (Figure 3.16).

Again, following the diagram, every path (starting at the root and ending with a terminal node) leads to the 1-node, so it is valid. On the other hand, it is not possible to

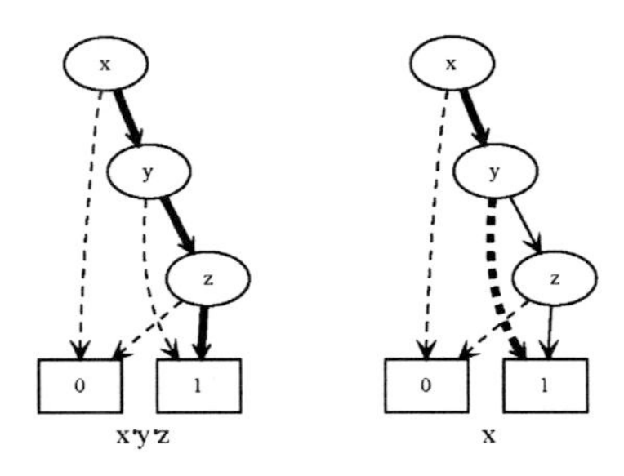

Figure 3.12: **Terms** of $x \cdot y \cdot z + x$

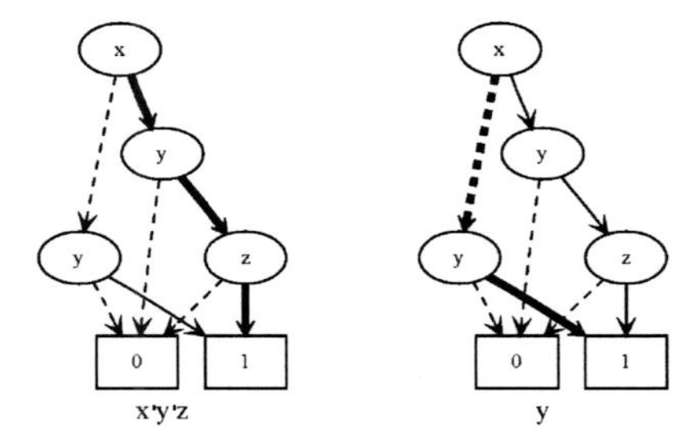

Figure 3.13: **Terms** of $x \cdot y \cdot z + y$

take a then-branch more than twice as the path reaches the terminal node at the latest after the second then-branch. The next variation of the polynomial will only contain terms of exact degree two: $a \cdot (b + c + d + e) + b \cdot (c + d + e) + c \cdot (d + e) + d \cdot e$ (Figure 3.17). The difference of the graphs is quite subtle. Two edges lead now to the 0-node which means that the corresponding path is not valid and the corresponding term does not occur in the polynomial. It is obvious to see that this makes exactly end those terminated paths at 0 which take only zero or one then-branch. The corresponding term of these paths has degree strictly less than two.

3.3 ZDDs and monomial orderings

Until now ZDDs have been a way to store polynomials. We have also identified terminated paths with terms in the polynomials. However, from computational algebra the terms

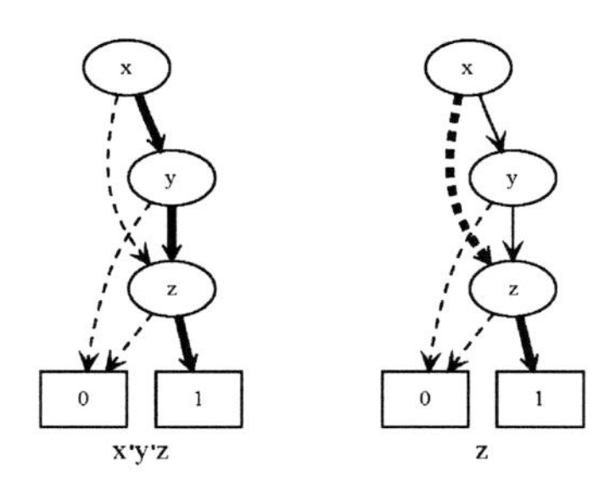

Figure 3.14: **Terms** of $x \cdot y \cdot z + z$

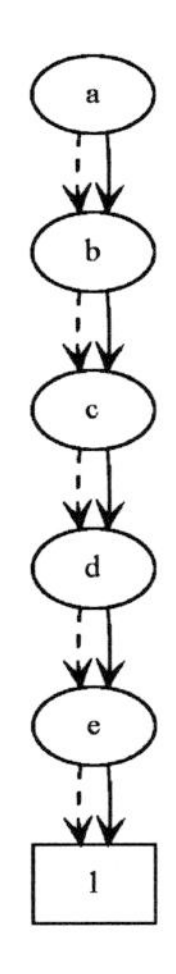

Figure 3.15: $(a + 1) \cdot (b + 1) \cdot (c + 1) \cdot (d + 1) \cdot (e + 1)$

of a polynomial are ordered by the monomial ordering. This monomial ordering is used in virtually every algorithm. One of the major advancements of this thesis lies in the treatment of monomial orderings on the level of ZDDs: We give iteration algorithms for the most important monomial orderings. Since the computation of the leading term is a special case of ordered iteration, algorithms for leading term computation can be derived.

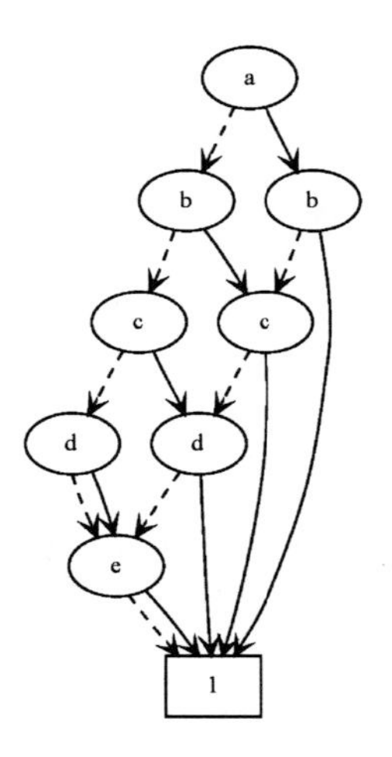

Figure 3.16: $a \cdot (b+c+d+e) + b \cdot (c+d+e) + c \cdot (d+e) + d \cdot e + a + b + c + d + e + 1$

3.3.1 Lexicographical ordering/natural path ordering

Definition 3.3.1. *We define a* ***natural ordering on valid paths*** *of a ZDD in the following way: Let* $b = (N_N, N_T, A_T, A_E, V, \text{var}, \text{vi})$ *be a ZDD,*

$$p_1 = (n_1, a_1, n_2, a_2 \dots, n_{k_0}, a'_{k_0} \dots n'_{k_1})$$

and

$$p_2 = (n_1, a_1, n_2, a_2 \dots, n_{k_0}, a''_{k_0} \dots n''_{k_1})$$

be valid paths with $a'_{k_0} \neq a''_{k_0}$. *Then we define* $p_1 > p_2$ *if* $a'_{k_0} \in A_T$.

The sequence of images in figure 3.18 illustrates the natural path ordering. Each picture highlights a path (ordered by the natural ordering which corresponds to the lexicographical monomial ordering).

In figure 3.19 a second less structured example is provided which shows the natural ordering of path: the polynomial $a \cdot b + a \cdot c \cdot d + a + b \cdot c \cdot d \cdot e + b \cdot c \cdot e + b$

Lemma 3.3.2. *No valid path is the beginning of another (valid path).*

Proof. Valid paths end at 1 which is a terminal node, so no origin of any arc. □

Remark 3.3.3. *Using notations from coding theory, lemma 3.3.2 is equivalent to saying that the set of valid paths in a diagram is a prefix code.*

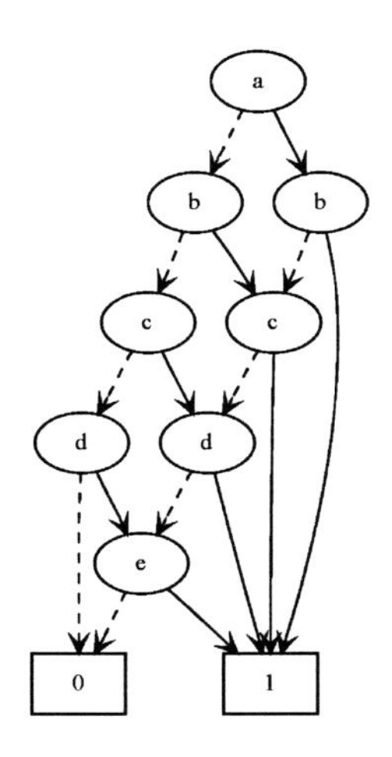

Figure 3.17: $a \cdot (b + c + d + e) + b \cdot (c + d + e) + c \cdot (d + e) + d \cdot e$

Corollary 3.3.4. *Let $P = \mathbb{Z}_2[x_1, \ldots, x_n]$, $x_i > x_j$ for $i < j$. Furthermore, we have the ZDD $b = (N_N, N_T, A_T, A_E, V, \mathrm{var}, \mathrm{vi})$. Any two distinct valid paths p_1, p_2 can be written in the following way:*

$$p_1 = (n_1, a_1, n_2, a_2 \ldots, n_{k_0}, a'_{k_0} \ldots n'_{k_1})$$

and

$$p_2 = (n_1, a_1, n_2, a_2 \ldots, n_{k_0}, a''_{k_0} \ldots n''_{k_1})$$

with

$$a'_{k_0} \neq a''_{k_0}.$$

Remark 3.3.5. *All distinct valid paths are comparable like in definition 3.3.1 since all of them start at the root of the diagram. Moreover $a'_{k_0} \in A_T$ of course means $a''_{k_0} \in A_E$ as there exists only one else-arc outgoing from a non-terminal node.*

Theorem 3.3.6. *Let $P = \mathbb{Z}_2[x_1, \ldots, x_n]$, $x_i > x_j$ for $i < j$. Furthermore, we have the ZDD $b = (N_N, N_T, A_T, A_E, V, \mathrm{var}, \mathrm{vi})$ and paths $p_1 = (n_1, a_1, n_2, a_2 \ldots, n_{k_0}, a'_{k_0} \ldots n'_{k_1})$ and $p_2 = (n_1, a_1, n_2, a_2 \ldots, n_{k_0}, a''_{k_0} \ldots n''_{k_1})$ (we may assume that $a'_{k_0} \neq a''_{k_0}$) valid paths satisfying $p_1 > p_2$. Then the lexicographical monomial ordering coincides with the natural ordering of paths.*

Proof. Let $m_1 = \mathrm{term}(p_1) = \prod_i x_i^{e_i}$, $m_2 = \mathrm{term}(p_2) = \prod_i x_i^{f_i}$. Then for $i < \mathrm{var}(n_{k_0})$ the exponents are equal:

$$e_i = f_i$$

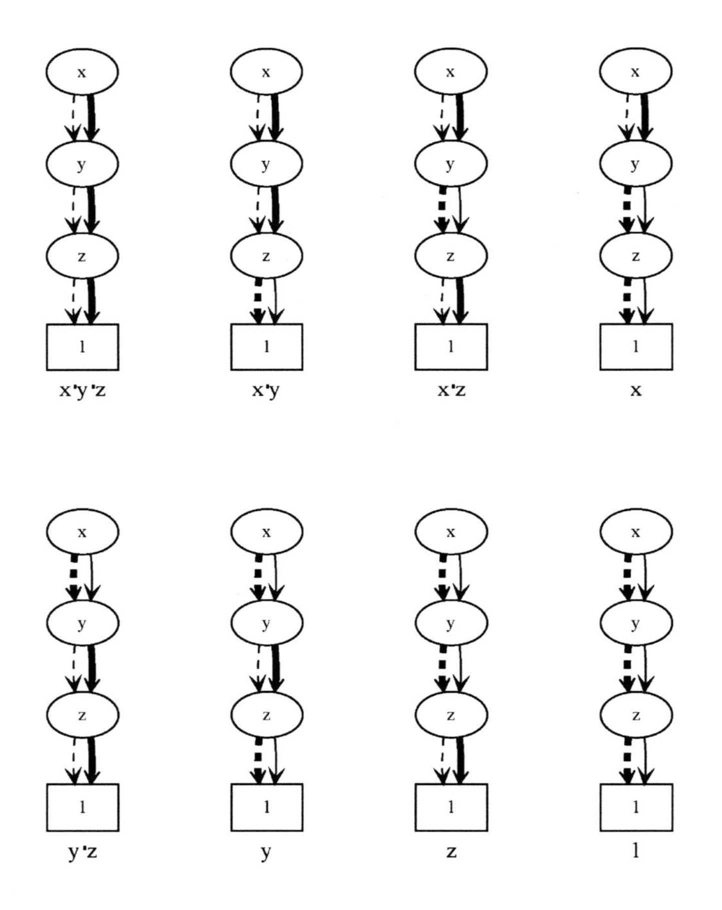

Figure 3.18: **Terms** of $(x+1)\cdot(y+1)\cdot(z+1)$ in Lexicographical Ordering

By definition 3.3.1 a'_{k_0} is a then-arc while a''_{k_0} is an else-arc. So x_i occurs only in m_1. As m_1 coincides with m_2 on the previous variables, for the lexicographical monomial ordering holds:

$$m_1 > m_2.$$

□

Corollary 3.3.7. *Let* $b = (N_N, N_T, A_T, A_E, V, \mathrm{var}, \mathrm{vi})$ *The natural ordering of valid paths*

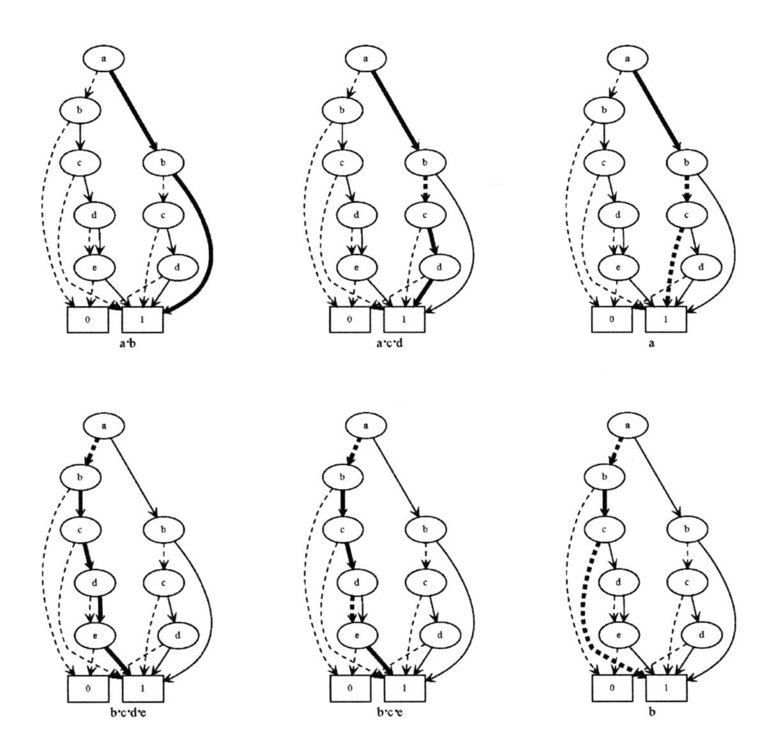

Figure 3.19: **Terms** of $a \cdot b + a \cdot c \cdot d + a + b \cdot c \cdot d \cdot e + b \cdot c \cdot e + b$ in Lexicographical Ordering

is an ordering relation on set(b).

Proof. Mapping a valid path p to term(p) is a bijection between the sets paths(b) and supp(polynomial(b)) and preserves the $>$-relation. □

It is challenging to present iteration algorithms in a simple, mathematical way as an iterator usually is seen as an object containing a state/position in the object that is iterated. However, it is still possible to describe the algorithms without introducing such kind of object oriented notion. For this purpose, in the following **generators** will be used. A **generator** is a subroutine which can return several values – not all at once (like returning a tuple), but each at a time (when it is requested). This will be denoted by the **yield** statement. We introduce this statement because its behaviour differs from a return statement: A generator runs until the next yield statement, returns that value and resumes there when it is called again.

The equivalence between natural ordering of paths and lexicographical monomial ordering (theorem 3.3.6) results directly in an iteration algorithm (algorithm 8): By definition 3.3.1 we only have to treat those originating from the then-branches first on each level. In all presented algorithms we will treat the special cases of constant ZDDs (0 or 1) first. The 0-ZDD has no valid paths while the 1-ZDD has an single (empty) valid path representing the monomial 1. After considering these special cases, we always can safely assume that the root nodes are non-terminal.

Algorithm 8 Lexicographically iterate terms: lex_iterate(p)

Input: Boolean polynomial p in ZDD form
Output: yields terms of p in lexicographical ordering

```
if p ≠ 0 then
  if (p = 1) then
    yield 1
  else
    /* p is not constant */
    for m ∈ lex_iterate(p^T) do
      yield topvar(p) · m
    for m ∈ lex_iterate(p^E) do
      yield m
```

In all these examples, the rightmost path consists only of then-branches and always leads to the terminal 1-node. While the position of the then-branches is a handy convention, the later might be surprising in the first. But it is direct consequence of definition 3.1.13. This can be formally proven in the following way.

Theorem 3.3.8. *Let* $b = (N_N, N_T, A_T, A_E, V, \mathrm{var}, \mathrm{vi})$ *and* $f = \mathrm{polynomial}(b) \notin \{0, 1\}$. *Then there exists a valid path* $p = (n_1, a_1, n_2, a_2 \ldots, n_k)$ $(n_k = 1)$*, where are all arc* a_i *are then-arcs. This path is the lexicographical leading term.*

Proof. Existence: This path p can be read of the diagram by starting at the root node and taking a then arc as often as possible. This can be done only finitely many times as the set of nodes is finite and the directed graph does not contain any directed cycles. So the last node n obtained in this way does not have a then-arc, so it is a terminal node (0 or 1). Since the graph does not represent a constant polynomial, the root node is not terminal and in this way we have constructed a non empty directed walk. In particular, there is a then-arc pointing to n. But as the diagram is reduced, we know that this then-arc does not lead to 0, so its terminus is 1 and the path is valid.

Leading term property: Let q be another valid path. So it starts at the same (root(b)) and coincides with p up to some node. By lemma 3.3.2 p does not terminate there, so it continues there with a then-arc. This means by definition 3.3.1 that $p > q$ which proves the claim. □

This explains why the lexicographical ordering works very natural for polynomials in ZDD representation. However, we have implemented quite sophisticated implementations of important other orderings (leading term calculation and ordered iteration) in PolyBoRi: degree lexicographical ordering, (ascending) degree reverse lexicographical ordering, block orderings (for degree ordered blocks). We will explain these iteration algorithms for these orderings in the next sections.

3.3.2 Degree orderings

In contrast to the natural path ordering we display the paths of the same polynomial now in degree-lexicographical ordering in figure 3.20.

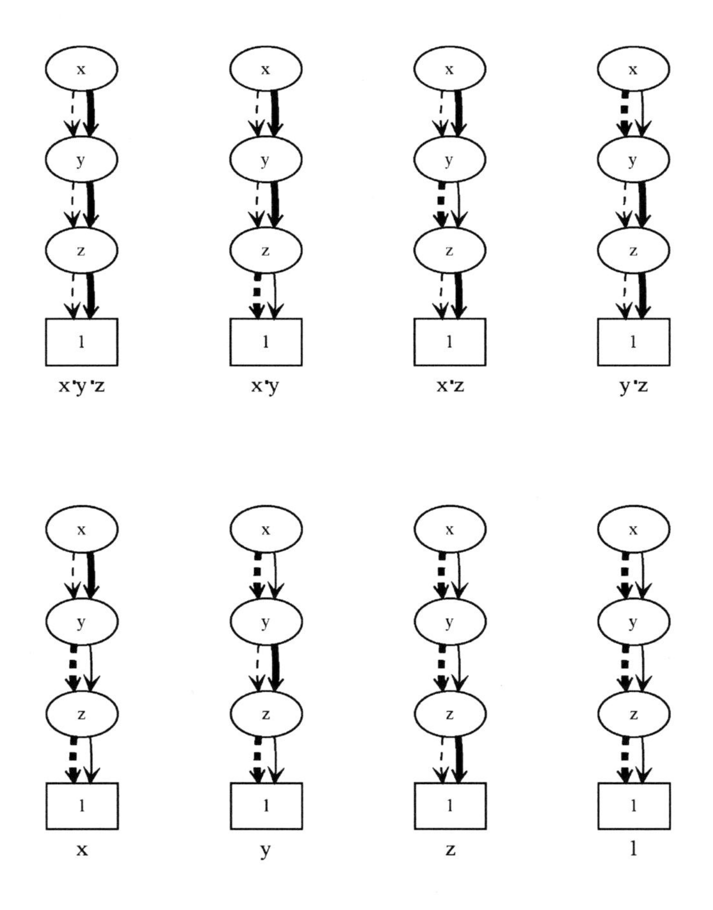

Figure 3.20: **Terms** of $(x + 1) \cdot (y + 1) \cdot (z + 1)$ in Degree Lexicographical Ordering

Also for the second example ($a \cdot b + a \cdot c \cdot d + a + b \cdot c \cdot d \cdot e + b \cdot c \cdot e + b$) we show the ordering of paths corresponding to degree-lexicographical ordering (figure 3.21).

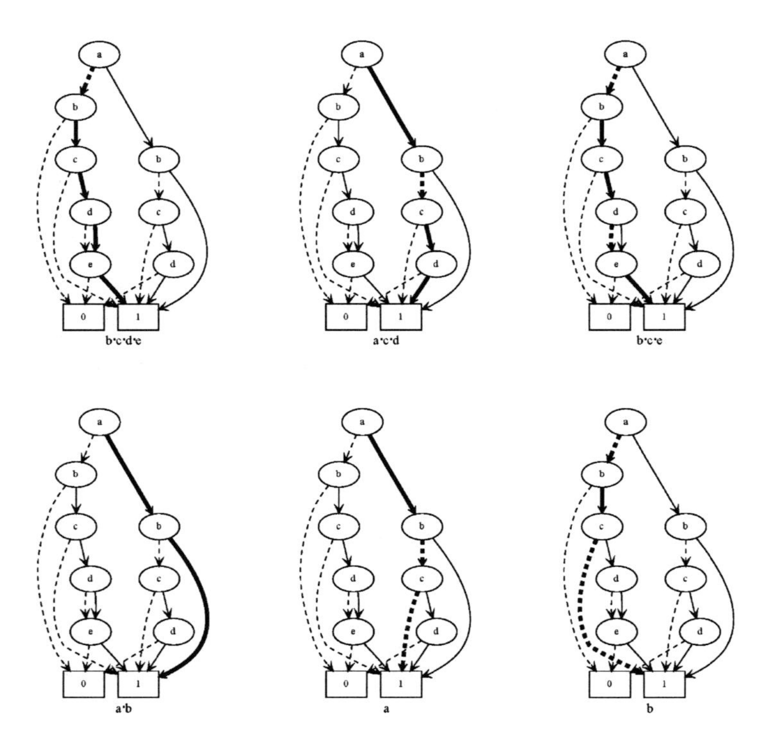

Figure 3.21: **Terms** of $a \cdot b + a \cdot c \cdot d + a + b \cdot c \cdot d \cdot e + b \cdot c \cdot e + b$ in Degree Lexicographical Ordering

The iterators implemented in PolyBoRi calculate successively the valid paths of a ZDD (terms of the polynomial). The calculation of the next path depends essentially on the current path. Moreover, in order to avoid conversion of the whole polynomial, they use only the ZDD structure of the polynomial. This is very important for performance if you iterate just over the first few terms.

Using this lexicographical iterator it is possible to implement a degree lexicographical iterator. A method for doing so is presented in algorithm 9: In each degree we can iterate lexicographically over the polynomial and check whether the term has the correct degree.

This algorithm can be optimized using an iterator which iterates lexicographically over a polynomial, but yields only the terms in a prescribed degree. So it does not to have to go too deep into the graph. This technique is presented in algorithm 10.

Algorithm 9 Degree Lexicographically iterate terms: deg_lex_iterate_simple(p)

Input: Boolean polynomial p in ZDD form
Output: yields terms of p in degree-lexicographical ordering

```
d := deg(p)
while d ≥ 0 do
   for m ∈ lex_iterate(p)  do
      if deg(m) = d then
         yield m
   d := d − 1
```

Algorithm 10 Lexicographically iterate terms in degree d: lex_iterate(p, d)

Input: Boolean polynomial p in ZDD form
Output: yields terms having degree d of p in lexicographical ordering

```
if p ≠ 0 and d ≥ 0 then
   if (p = 1) then
      if d = 0 then
         yield 1
   else
      /* p is not constant */
      for m ∈ lex_iterate(p^T, d − 1)  do
         yield topvar(p) · m
      for m ∈ lex_iterate(p^E, d)  do
         yield m
```

Switching the order of branches in algorithm 10 it is possible to iterate over a polynomials terms in degree d in reverse lexicographical ordering. This is shown in algorithm 11.

Algorithm 11 Reverse Lexicographically iterate terms in degree d: rev_lex_iterate(p, d)

Input: Boolean polynomial p in ZDD form
Output: yields terms having degree d of p in lexicographical ordering

```
if p ≠ 0 and d ≥ 0 then
  if (p = 1) then
    if d = 0 then
      yield 1
  else
    /* p is not constant */
    for m ∈ rev_lex_iterate(p^E, d) do
      yield m
    for m ∈ rev_lex_iterate(p^T, d − 1) do
      yield topvar(p) · m
```

Finally, this can be used to construct algorithm 12 for iteration in degree reverse lexicographical ordering.

Algorithm 12 Degree Lexicographically iterate terms: deg_rev_lex_iterate

Input: Boolean polynomial p in ZDD form
Output: yields terms of p in degree-lexicographical ordering

```
d := deg(p)
while d ≥ 0 do
  for m ∈ rev_lex_iterate(p, d) do
    yield m
  d := d − 1
```

3.3.3 Block orderings

The implementation of block orderings using ZDDs is very difficult as it involves all problems of the orderings inside the blocks, together with the challenge of composing them to efficient algorithms for leading term calculation and iteration. However, the basic principle, how such an ordering can be implemented, can be described quite naturally.

Definition 3.3.9. *Given the polynomial ring $\mathbb{Z}_2[x_1, \ldots, x_n]$ a* ***block ordering*** $>$ *with two blocks is given by an integer $1 < i_0 \leq n$ and monomial orderings $>_1$ on $\mathbb{Z}_2[x_1, \ldots, x_{i_0-1}]$ and $>_2$ on $\mathbb{Z}_2[x_{i_0}, \ldots, x_n]$. We define an ordering to be a block ordering composed of $>_1$, $>_2$, if for all monomial $m_1, m_2 \in \mathbb{Z}_2[x_1, \ldots, x_{i_0-1}], n_1, n_2 \in \mathbb{Z}_2[x_{i_0}, \ldots, x_n]$, we have:*

$$m_1 \cdot m_2 > m_2 \cdot n_2 \iff m_1 >_1 m_2 \text{ or } (m_1 = m_2 \text{ and } n_1 >_2 n_2).$$

We restrict ourself to two blocks here, as ordering consisting of arbitrary many blocks can be constructed recursively.

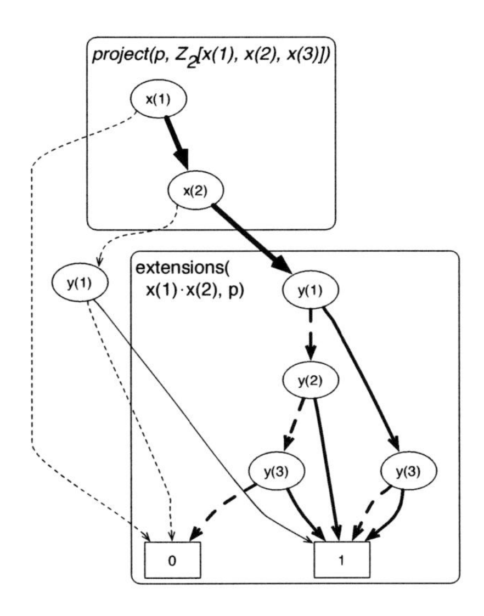

Figure 3.22: Two blocks of variables.

We illustrate the idea in Figure 3.22. The image shows the complete polynomial. The upper box shows the part of the ZDD which belongs to the first block. Each path in this ZDD leading to a node in the second block can be extended to one or several valid paths. Each valid path in the original diagram can be projected to a path in this upper box. We mark every path corresponding to a monomial which is a multiple of $x_1 \cdot x_2$ bold in diagram. The lower box contains the endings of these paths. We call it the extension subdiagram of $x_1 \cdot x_2$. Iteration over a polynomial in the block ordering means: Iterating over the paths in the projection and for each path in the projection iterating over the paths in the extension subdiagram in the last block.

We formalize this in the following:

Definition 3.3.10. *Let p be a Boolean polynomial. A ZDD p′ is called the* ***projection*** *into* $\mathbb{Z}_2[x_1, \ldots, x_{i_0-1}]$ *(*$p' = \mathrm{project}(p, \mathbb{Z}_2[x_1, \ldots, x_{i_0-1}])$*) if the following holds:*

$$\mathrm{polynomial}(p') = \{m \in \mathbb{Z}_2[x_1, \ldots, x_{i_0-1}] | \exists n \in \mathbb{Z}_2[x_{i_0}, \ldots x_n] : m \cdot n \in \mathrm{supp}(p)\}$$

For a polynomial represented in ZDD form, the ZDD of the projection in the first is generated by replacing all edges which link to nodes from the first block to those of the second by direct connections to the terminal 1-node. For $m \in \mathrm{project}(p, \mathbb{Z}_2[x_1, \ldots, x_{i_0-1}])$ *the set*

$$\mathrm{extensions}(m, p) := \{n \in \mathbb{Z}_2[x_{i_0}, \ldots x_n] : m \cdot n \in \mathrm{supp}(p)\}$$

corresponds to a subdiagram of the ZDD (see Figure 3.22).

Remark 3.3.11. *Note that definition 3.3.10 is given in algebraic terms. An implementation will of course use the ZDD structure. This has been described informally above. Doing that formally would involve a lot of machinery and not provide additional insights.*

We combine these results in the algorithm 13. The algorithm is trivially correct. However, as explained above and visualized in Figure 3.22, since $\text{project}(p, \mathbb{Z}_2[x_1, \ldots, x_{i_0-1}])$ and $\text{extensions}(m, p)$ can be read off the diagram structure, they do not have to be computed. The essential part of this algorithm is that it reduces the problem to iteration in the two blocks. Of course it is very challenging to implement it as for example $\text{project}(p, \mathbb{Z}_2[x_1, \ldots, x_{i_0-1}])$ is not really a subdiagram because it corresponds to a part of the ZDD that does not end at a terminal node. Also the iteration over it has to be simulated without constructing it.

Algorithm 13 iterate terms: blockIterate

Input: Boolean polynomial p in ZDD form
Output: yields terms of p in block ordering
  for $m \in \text{iterator}_{>_1}(\text{project}(p, \mathbb{Z}_2[x_1, \ldots, x_{i_0-1}]))$ **do**
    for $n \in \text{iterator}_{>_2}(\text{extensions}(m, p))$ **do**
      yield $m \cdot n$

3.3.4 Implementation of iterators in PolyBoRi

In PolyBoRi the iterators are implemented using C++. There exists no generator in C++: Iterators are objects containing a method for reading out the current value and a method for increasing the iterator (progressing to the next term). The state of the generator function can be saved in the generator object. It involves all its local variables. Since the generator functions are called recursively, a complete stack of variables has to be stored in the iterator object. When we look a little bit closer at it, the only data needed to store in a stack are the nodes of a directed walk.

3.4 Dependence of ZDDs on the order of variables

It is quite obvious that the structure of the ZDDs depends on the ordering of variables: In the trivial cases just some nodes are switched. However, in general the ZDD can look very different.

We start with a simple example:

The polynomial $a \cdot b + a \cdot c + a + 1$. In figure 3.23 it is shown in two different variable orderings. One can see that the total number of nodes varies for the two orderings. The ZDD using alphabetical variable ordering has four nodes (three non-terminal nodes and one terminal nodes) while the graph in reversed alphabetical variable ordering has six nodes. Moreover it can be observed that using the first ordering the 0-terminal node does not occur in the graph.

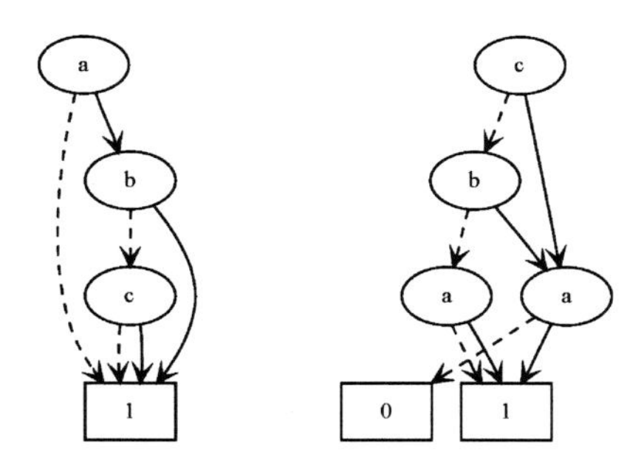

Figure 3.23: $a \cdot b + a \cdot c + a + 1$, normal and reversed order

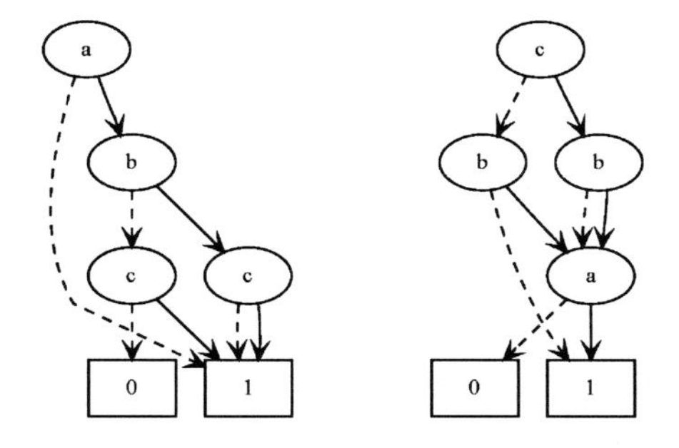

Figure 3.24: $a \cdot b \cdot c + a \cdot b + a \cdot c + 1$, alphabetical and reversed variable order

In the figure 3.24 a slightly more complex example is shown: The polynomial $a \cdot b \cdot c + a \cdot b + a \cdot c + 1$

While the differences above were in expected magnitude and the polynomials looked quite artificial, we continue with an example corresponding to the highest carry bit of an 3-bit adder. For this case of 3 bits the highest order carry bit has the formula: $a_2 \cdot b_2 + a_2 \cdot a_1 \cdot b_1 + a_2 \cdot a_1 \cdot a_0 \cdot b_0 + a_2 \cdot b_1 \cdot a_0 \cdot b_0 + b_2 \cdot a_1 \cdot b_1 + b_2 \cdot a_1 \cdot a_0 \cdot b_0 + b_2 \cdot b_1 \cdot a_0 \cdot s$. First the ZDD for the naïve alphabetical ordering of variables is presented in figure 3.25(a). It looks quite huge and not very structured. But if we optimize the ordering of variables, the graph is shown in figure 3.25(b).

Increasing the number of bits, the number of terms will grow exponentially while the number of nodes in the optimized ZDD-representation just increases linearly. Using this

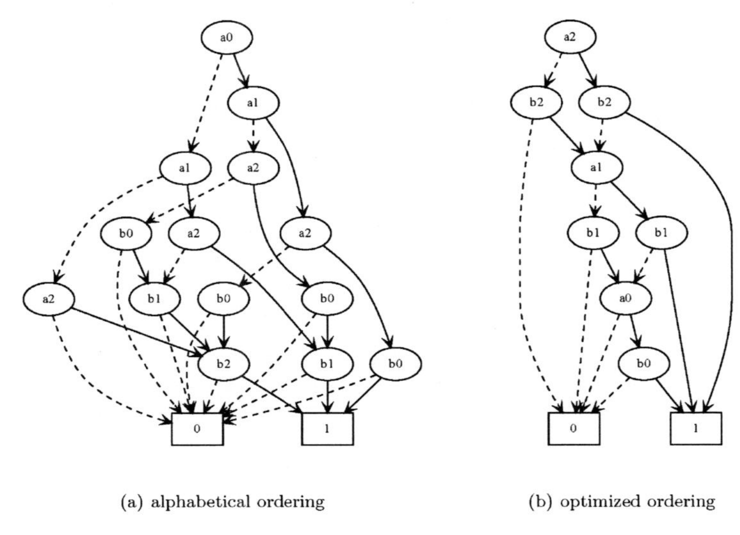

(a) alphabetical ordering

(b) optimized ordering

Figure 3.25: Carry bit of a 3-bit adder

ordering of variables makes also the polynomial computations much faster.

Chapter 4

Algorithmic optimizations using ZDD

This chapter shows how to optimize algorithms used for computing Gröbner bases computations over ZDD-based data structures. We start with basic arithmetic, continue with two for computing normal forms, and finally we present some possibilities for applying ZDDs to the management of critical pairs in the Buchberger algorithm.

4.1 Arithmetic

This section is devoted to addition and multiplication as well as good algorithms for them using ZDD data structures.

Addition and multiplication form the very basic operations in commutative algebra. However, for performant implementations, it is often necessary to find different approaches which are not motivated purely algebraically. We illustrate this in many examples in section 7.2. When computing with ZDDs, the usual paradigm is to use set operations as often as possible:

Since terms in Boolean polynomial do not have any coefficients, a Boolean polynomial $p \in \mathbb{B}$ can be identified with its support set $\operatorname{supp}(p) \in \mathcal{P}(\{x^\alpha | \alpha \in \mathbb{N}^n\})$ where $p = \sum_{s \in \operatorname{supp}(p)} s$. Considering this set representation, even the most basic algebraic operations addition and multiplication can be expressed in terms of set operations.

4.1.1 Addition

The goal of this section is to formulate addition of Boolean polynomials as set operation and to provide a recursive algorithm which computes the sum of two Boolean polynomials. Addition of Boolean polynomials is equivalent to the symmetric difference of their support sets:

$$\operatorname{supp}(p+q) = (\operatorname{supp}(p) \cup \operatorname{supp}(q)) \setminus (\operatorname{supp}(p) \cap \operatorname{supp}(q)).$$

Since ZDD representation of a Boolean polynomial can be interpreted as the Boolean polynomial itself or its support set, the operation on the support set is just what we need. However, we formulated the basic addition using three set operations: union, complement, and intersection. These are already available as basic ZDD operations in the CUDD library.

For practical applications it is appropriate to provide a direct implementation of the symmetric difference which avoids unnecessary intermediate results.

Algorithm 14 shows a recursive approach for such an addition. For the presentation of the algorithm, we would like to repeat, that f^T and f^E denote the then- and else-branch of a ZDD (see definition 3.1.15).

Algorithm 14 Recursive addition $f + g$

Input: $f, g \in \mathbb{B}$
  if $f = 0$ **then**
    $h = g$
  else if $g = 0$ **then**
    return f
  else if $f = g$ **then**
    return 0
  else
    set $x_\nu = \text{top}(f)$, $x_\mu = \text{top}(g)$
    if $\nu < \mu$ **then**
      return $\text{ite}(x_\nu, f^T, f^E + g)$
    else if $\nu > \mu$ **then**
      return $\text{ite}(x_\mu, g^T, f + g^E)$
    else
      return $\text{ite}(x_\nu, f^T + g^T, f^E + g^E)$

The addition and multiplication algorithms are fully cacheable: An implementation should make sure, that these are computed only once for a given pair of inputs. After that the result is stored in an internal cache structure. The following calls can check the cache before the computation and retrieve the result. We do not include the caching into the pseudo-code, as it follows the same principles everywhere and can be seen independent from the actual recursive algorithms. The lookup can be implemented cheaply because polynomials have a unique representation as ZDDs. Hence, previous computations of the sums of the form $f+g$ can be reused. The advantage of a recursive formulation is, that this also applies to those subpolynomials which are generated by f^T and f^E. It is very likely, that common subexpressions can be reused during Gröbner base computation because of the recurring multiplication and addition operations which are used in Buchberger-based algorithms for elimination of leading terms and the tail-reduction process.

4.1.2 Multiplication

The Boolean multiplication in algorithm 15 follows the same approach. Note, it does not compute the normal product $(\cdot)$, but the Boolean product ($\star$, see definition 2.1.3) which is reduced against the field polynomials. This is indeed necessary, as we do not represent general polynomials by ZDDs and the normal product of two Boolean polynomials might not be a Boolean polynomial any more. However, if variables of right- and left-hand side polynomials are distinct, both operations coincide. On the other hand, we would like to note, that it is very practical being able to use Boolean multiplication as in most applications we prefer polynomials reduced against field equations. Moreover, the Boolean product has at most as many terms as the normal product and often much less.

Algorithm 15 Recursive multiplication $f \star g$

Input: $f, g \in \mathbb{B}$
  if $f = 1$ **then**
    return g
  else if $f = 0$ or $g = 0$ **then**
    return 0
  else if $g = 1$ or $f = g$ **then**
    return f
  else
    $x_\nu = \text{top}(f)$, $x_\mu = \text{top}(g)$
    if $\nu < \mu$ **then**
      set $p_1 = f^T$, $p_0 = f^E$, $q_1 = g$, $q_0 = 0$
    else if $\nu > \mu$ **then**
      set $p_1 = g^T$, $p_0 = g^E$, $q_1 = f$, $q_0 = 0$
    else
      set $p_1 = f^T$, $p_0 = f^E$, $q_1 = g^T$, $q_0 = g^E$
    return $\text{ite}(x_{\min(\nu,\mu)}, (p_0 + p_1)\star q_1 + p_1 \star q_0, p_0 \star q_0)$

4.1.3 Fast multiplication

In algorithm 16 we repeat the Karatsuba's univariate fast multiplication, as given in [J. von zur Gathen and J. Gerhard, 1999]. The number of multiplications is decreased by

Algorithm 16 fast univariate multiplication $f \cdot g$

Input: $f, g \in K[x]$ univariate polynomials
  $r = \min\{r \in \mathbb{N} | 2^r > \max(\deg(f), \deg(g))\}$
  if $r = 0$ **then**
    return $f \cdot g$ /* normal field multiplication */
  set $b = 2^{r-1}$
  $f_0 = f \mod x^b$
  $g_0 = g \mod x^b$
  $f_1 = f / x^b$
  $g_1 = g / x^b$
  set $l = f_0 \cdot g_0$
  set $h = f_1 \cdot g_1$
  return $h \cdot x^{2^r} + ((f_0 + f_1) \cdot (g_0 + g_1) - l - h) \cdot x^b + l$

reusing $h = f_1 \cdot g_1$ and $l = f_0 \cdot g_0$ and building the product $(f_0 + f_1) \cdot (g_0 + g_1)$. Based on the ideas of Karatsubas's algorithm for univariate fast multiplication, we present a fast Boolean multiplication algorithm for multivariate Boolean polynomials. We have never found literature reference for multivariate fast multiplication. But is obvious, that it is possible, when choosing a main variable x in each step and considering the univariate polynomial ring in x over the polynomial ring in the over variables. This choice of course involves heuristic. When using ZDDs, there exists an obvious heuristic which should be the main variable: The variable of the root nodes of the diagrams (when both diagrams have the same root node, which can be assumed without loss of generality).

Algorithm 17 Fast multiplication $h = f \star g$

Input: $f, g \in \mathbb{B}$
  if $f = 1$ **then**
    return g
  else if $f = 0$ or $g = 0$ **then**
    return 0
  else if $g = 1$ or $f = g$ **then**
    return f
  else
    $x_\nu = \operatorname{top}(f)$, $x_\mu = \operatorname{top}(g)$
    if $\nu \neq \mu$ **then**
      if $\nu < \mu$ **then**
        set $p_1 = f^T$, $p_0 = f^E$, $q_1 = g$, $q_0 = 0$
      else if $\nu > \mu$ **then**
        set $p_1 = g^T$, $p_0 = g^E$, $q_1 = f$, $q_0 = 0$
      return $\operatorname{ite}(x_{\min(\nu,\mu)}, p_0 \star q_1 + p_1 \star q_1 + p_1 \star q_0, p_0 \star q_0)$
    else
      set $e = f^E \star g^E$
      return $\operatorname{ite}(x_\nu, (f^T + f^E) \star (g^T + g^E) - e, e)$

While algorithm 15 uses three recursive multiplication calls (in the interesting case $\nu = \mu$), the asymptotically fast algorithm algorithm 17 uses only two. It might be surprising, that the classical univariate fast multiplication algorithm has three recursive calls which is one more than our fast multiplication. We were able to reduce the number of calls using the quotient ring structure. The key is some tricky scheme (using $x = x_\nu$):

$$\begin{aligned} f \star g &= \operatorname{ite}(x, f^T \star g^T + f^T \star g^E + f^E \star g^T, f^E \star g^E) \\ &= \operatorname{ite}(x, (f^T + f^E) \star \operatorname{ite}(x, g^T + g^E) - f^E \star g^E, f^E \star g^E) \end{aligned}$$

The obvious disadvantage of fast multiplication is the hidden density assumption: Counting the recursive multiplication calls is only a measure for performance, when this calls all take a similar time. However, $(f^T + f^E) \star (g^T + g^E)$ might be quite hard to compute compared to the multiplications in algorithm 15, as the summation can make the polynomials more dense. This can be observed in Table 4.1.3. One should pay attention to that fact, that already the classical multiplication is very fast in our formulation suitable for the ZDD data structures (algorithm 15). We always multiplied two polynomials of equal number of terms. We always denote that size by its logarithm to the base 2, so in the biggest example we multiply two polynomials, each with $2^{18} = 262144$ terms. We tried the first example with 16 variables and 2^{14} terms with SINGULAR 3-1-0. However, it took already 5 minutes for the multiplication. Moreover the result is not reduced against the field ideal. The reduction of the multiplication result against the quotient ideal took nearly two hours. Hence, even while the table lists timings for single multiplications, the dimension of the examples is quite large.

variables	$\log_2$ terms per factor	time classical mult. in s	time fast mult. in s
16	14	3.00	0.09
18	16	27.96	0.42
18	13	13.80	0.40
18	7	0.07	0.05
21	18	793.16	5.13
21	16	498.68	4.20
21	15	311.31	3.77
21	14	157.37	3.21
21	13	70.26	2.68
21	11	9.62	1.60
21	5	0.01	0.03

Table 4.1: Boolean multiplication: two recursive algorithms using set operations, one oriented on the classical multiplication, the other is the new fast algorithm for multiplication of Boolean polynomials.

So algorithm 15 remains the default multiplication strategy in PolyBoRi and algorithm 17 can be used as supplement for dense cases. Actually, this assumption of density is shared by other asymptotical fast methods (matrix multiplication, univariate polynomial multiplication). It will be a challenge in future to provide fast heuristics, allowing to switch between classical and these asymptotically fast methods.

4.2 Normal form against a set of monomials

In this section we give a first, easy example of ZDD-style recursive normal form computations for special systems of generators. Consider a Boolean polynomial f and a set of (Boolean) monomials G. Then we have $g = \mathrm{REDNF}(f, G)$ if and only if $g \in f + \langle G \rangle$. Note, that it does not matter in this case, if we add the field polynomials to G.

We can identify the set G with a single polynomial where exactly the elements of G appear with coefficient 1. So, both f and G can be represented as ZDDs. Hence, we will formulate the procedure recursively in the language of ZDDs, but for the sake of simplicity, we skip caching in the pseudo-code.

Computing the normal form of a polynomial f against a set of monomials G, means just cancelling all terms in f which are divisible by an element of G. Algorithm 18 does it very efficiently (properly implemented) and it is the easiest example for directly computing the reduced normal form recursively without the classical Buchberger normal form algorithm.

This algorithm is small and useful. As it is extreme fast, it can be used to speed up general normal form algorithms, where a sufficient part of the generating system consists

Algorithm 18 Normal form against a set of monomials

Input: f a polynomial, G a set of monomials, both encoded as ZDD.
Output: nf_mon_set(f, G) = REDNF(f, G)

```
if 1 ∈ G then
  return 0
if f ∈ {0, 1} then
  return f
while top(f) > top(G) do
  G = G^E
if G = ∅ then
  return f
/* monomial in which topvar(f) occurs must be tested for divisibility with the then
and the else-branch of G */
if top(f) = top(G) then
  return ite(top(f),
    nf_mon_set(nf_mon_set(f^T, G^E), G^T),
    nf_mon_set(f^E, G^E))
else
  return ite(top(f), nf_mon_set(f^T, G),
    nf_mon_set(f^E, G))
```

of monomials. Moreover many subproblems in Gröbner bases computations somehow involve this operation as can be seen in section 4.4.2.

4.3 Linear lexicographical lead reductions

In applications like verification of digital circuits and cryptography, it is often important to calculate lexicographical normal forms against a system $F = \{f_i | i \in \{1, \ldots, m\}\}$, for some integer $m \leq n$, of Boolean polynomials with pairwise different linear lexicographical leading terms $l_i = \mathrm{lm}(f_i)$:

$$\deg(l_i) = 1 : \exists j_i : l_i = x_{j_i}$$

We call F a **linear lexicographical lead rewriting system.**

Theorem 4.3.1. *A linear lexicographical lead rewriting system is a lexicographical Boolean Gröbner basis.*

Proof. We have to show that each pair (f, g) $(f, g \in F \cup \{x_1^2 + x_1, \ldots, x_n^2 + x_n\}$ of polynomials has a standard representation. Since the product criterion applies otherwise, we can assume that f has a linear lexicographical leading term and g is a field polynomial. As all field polynomials are included in our system and the leading term of f, we can apply theorem 2.2.10 to see that the pair has a nontrivial t-representation. □

The normal form computation against F can be done in a recursive ZDD computation. While for calculating general normal forms such formulations probably do not exist, we can present algorithms for lexicographical normal forms against systems of Boolean polynomials with linear lexicographical leading terms. For the performance of the PolyBoRi framework it was an essential step to find these algorithms.

The first problem we encounter is that recursive functions in CUDD take at most three arguments, but F may contain arbitrary many arguments.

So we need to encode all elements of F into a single ZDD. This can be done in the following way.

To simplify the presentation of the algorithm, we will assume that $i_j < i_{j+1}$ for all $j \in \{1, \ldots, m-1\}$.

We encode F by $\mathrm{ite}(x_{i_1}, \mathrm{ite}(x_{i_2}, \ldots \mathrm{ite}(x_{i_m}, 1, \mathrm{tail}(f_m)) \ldots, \mathrm{tail}(f_2)), \mathrm{tail}(f_1))$. This is visualized in Figure 4.1.

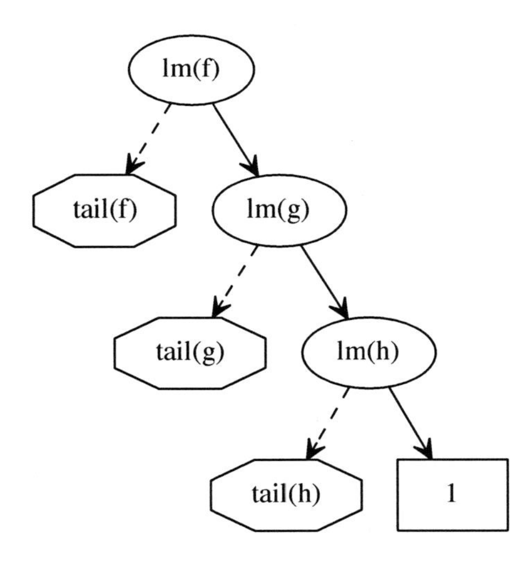

Figure 4.1: Encoding of system of polynomials with linear, lexicographical leading terms

In this way, we can iterate over the leading terms of F by always following the then-branch. The elements of F can be reconstructed by accessing the respective else-branch.

For this encoding of F we define the reduction algorithm for a Boolean polynomial against $F \cup \{x_1^2 + x_1, \ldots, x_n^2 + x_n\}$ in algorithm 19 (llnf).

There exists a variant of algorithm 19, which we call llnf*. It replaces the last return-statement

$$\mathrm{llnf}(F^E, F^T) \star \mathrm{llnf}(p^T, F^T) + \mathrm{llnf}(p^E, F^T)$$

Algorithm 19 llnf: lexicographical normal form against systems with linear leads

Input: p Boolean polynomial, F system of Boolean polynomials with pairwise different linear lexicographical leading terms encoded as described above
Output: result $\mathrm{REDNF}(p, F \cup \{x_1^2 + x_1, \ldots, x_n^2 + x_n\})$
  if $F = \{1\}$ **then**
    return p
  if $p \in \{0, 1\}$ **then**
    return p
  if $\mathrm{top}(p) > \mathrm{top}(F)$ **then**
    return $\mathrm{llnf}(p, F^T)$
  if $\mathrm{top}(p) < \mathrm{top}(F)$ **then**
    return $\mathrm{topvar}(F) \star \mathrm{llnf}(p^T, F) + \mathrm{llnf}(p^E, F)$
  /* now we know $\mathrm{top}(p) = \mathrm{top}(F)$ */
  return $\mathrm{llnf}(F^E, F^T) \star \mathrm{llnf}(p^T, F^T) + \mathrm{llnf}(p^E, F^T)$

by
$$\mathrm{llnf}^*(F^E \star p^T + p^E, F^T).$$
In the case of equal top variables this second variant recursively calls itself only once, instead of three times. Actually llnf^* is can also be seen as a specialization of algorithm 6 for the case of linear lexicographical lead reductions: It contains no choices and operates directly on the diagrams.

Theorem 4.3.2. *Let F be as stated above, p a Boolean polynomial, then algorithm 19 terminates and returns a reduced normal form of p against $F \cup \{x_1^2 + x_1, \ldots, x_n^2 + x_n\}$.*

Proof. For constant polynomials or the empty set the algorithm is obviously correct. Let $\deg(p) > 0$ and $F \neq \emptyset$. Since the recursive calls terminate (see Remark 3.1.16), the algorithm terminates by induction on $\min(\mathrm{top}(p), \mathrm{top}(F))$. We have to show: $\mathrm{llnf}(p, F)$ is reduced against $F \cup \{x_1^2 + x_1, \ldots, x_n^2 + x_n\}$ and lies in the same residue class like p modulo $F, \{x_1^2 + x_1, \ldots, x_n^2 + x_n\}$. Without loss of generality we assume $\mathrm{top}(p) = \mathrm{top}(F) = j_1$. We can further assume by induction that the theorem holds for the recursive calls. Therefore,

$$\begin{aligned} \mathrm{llnf}(F^E, F^T) \star \mathrm{llnf}(p^T, F^T) + \mathrm{llnf}(p^E, F^T) &= \\ \mathrm{llnf}(\mathrm{tail}(f_1), \{f_2, \ldots, f_m\}) \star \mathrm{llnf}(p^T, \{f_2, \ldots, f_m\}) + & \\ \mathrm{llnf}(p^E, \{f_2, \ldots, f_m\}) &\equiv (\mod \{f_2, \ldots, f_m, \\ & \quad x_1^2 + x_1, \ldots, x_n^2 + x_n\}) \\ x_{j_1} \cdot p^T + (p^E) &= p \end{aligned}$$

Hence, the result lies in the same residue class. Now, we show that it is reduced:

Since the recursive call return polynomials, which are reduced against $\{x_1^2 + x_1, \ldots, x_n^2 + x_n\}$, they are Boolean polynomials and their addition and Boolean multiplication yields also Boolean polynomials (so reduced against $\{x_1^2 + x_1, \ldots, x_n^2 + x_n\}$). It remains to show that the result of the algorithm is reduced against F. In a similar way, we use the fact that the recursive call returns normal forms of F^E, p^T, p^E against
$$\{f_2, \ldots, f_m, x_1^2 + x_1, \ldots, x_n^2 + x_n\}.$$

A polynomial p is reduced against $\{f_2, \ldots, f_m\}$, if and only if it does not involve any x_{j_k} for $k > 1$. So the sum and (Boolean) product of two such polynomials are again reduced against F. Since the arguments to the recursive calls do not involve any variable x_k with $k \leq j_1$, their reduced lexicographical normal form against $\{f_2, \ldots, f_m\} \cup \{x_1^2 + x_1, \ldots, x_n^2 + x_n\}$ does not involve such variables either. In particular, it does not involve $x_{j_1} = \mathrm{lm}(f_1)$. Hence, these recursive results are actually reduced against

$$F \cup \{x_1^2 + x_1, \ldots, x_n^2 + x_n\}$$

and do not involve any x_{j_1} as well as their (Boolean) product and sum. □

In the case, that F is not only a Boolean Gröbner basis, but also reduced, we can simplify the algorithm leaving out the reduction of the else branches of F.

Algorithm 20 llnfredsb: lexicographical normal form against systems with linear leads

Input: p Boolean polynomial, F reduced Boolean Gröbner basis, encoded as above
Output: result llnfredsb$(p, F \cup \{x_1^2 + x_1, \ldots, x_n^2 + x_n\})$
  if $F = \{1\}$ **then**
    return p
  if $p \in \{0, 1\}$ **then**
    return p
  if top$(p) >$ top(F) **then**
    return llnfredsb(p, F^T)
  if top$(p) <$ top(F) **then**
    return topvar$(F)\star$ llnfredsb(p^T, F) + llnfredsb(p^E, F)
  /* now we know top$(p) =$ top(F) */
  return $F^E\star$ llnfredsb(p^T, F^T) + llnfredsb(p^E, F^T)

The results presented in Figure 4.2 show that choosing the right reduction strategy is quite difficult. Moreover, these big differences provide insight in the enormous importance of heuristics and a wide variety of functions specialized on different kinds of polynomial systems.

The used machine is a Macbook Pro with a 2.5 GHz Intel Core 2 Duo and 4GB RAM. Each example is run with the normal Buchberger algorithm (which does essentially the elimination and some little extra computation to achieve the Gröbner basis in the remaining variables). The timings for algorithm 20 also include the time needed to reduce the system F using the same reduction method. Furthermore we reordered the ring variables and computed all three linear lexicographical lead reduction algorithms a second time. Here an optimized variable ordering was generated, which preserves the leading terms of the system. This is described in more detail in section 5.2.2.

We can see in both cryptographic examples (AES, CTC) that the optimized variable ordering does not seem to improve the situation. This was expected, as the examples are very regularly structured and the natural ordering of variables is very near the setting, the optimizing function computes. It does not seem to be obvious, whether to prefer the variant computing a reduced Gröbner basis of F first. But clearly, llnf* performs usually much better using an optimized topological ordering.

	simple topological ordering				optimized var. order.		
	Buchberger	llnf	llnf*	llnfredsb	llnf	llnf*	llnfredsb
mult 8x8	0.65	3.27	12.19	3.19	0.66	0.09	0.53
mult 10x10	15	891.92	>7200	627.46	891	0.32	907
mult 16x16						76.05	
viscoherencep5	1.5	0.075	0.076	0.18	0.29	0.29	0.32
aes 10 1 2 4	4.3	1022.85	813	0.17	1112	3636	0.27
ctc Nr3 B5 n20	73	13.05	6.2	13.03	45	9.26	22.51

Table 4.2: Linear lexicographical lead reduction examples from formal verification and cryptanalysis. All times are measured in seconds.

Overall, these enormous differences show the huge potential of a variety of optimized solutions together with a good heuristic.

The multiplier example are classical multiplier designs as given in the text books [Patterson and Hennessy, 2009] and [Koren, 2001]. They were provided by Wedler [2007]. The CTC [Courtois, 2006] example is due to Albrecht [2006]. The viscoherencep example is part of the AIGER distribution [Biere, 2007]. The AES example was provided by Stanislav Bulygin. We have formed several series [Bulygin and Brickenstein, 2010] of algebraic attacks on small scale AES/SR ([Cid et al., 2005]). The reductions presented in this section form an essential algorithmic ingredient of these attacks. Although the examples form all problems which are solvable with reductions against a linear lexicographical lead system, they have quite different characteristics. In [Brickenstein et al., 2009] it has been shown that classical computer algebra systems like Magma [Bosma et al., 1997] and SINGULAR [Greuel et al., 2001, 2009] are not able to tackle even the 8-bit multiplier in reasonable time. In [Brickenstein and Dreyer, 2009a], we have demonstrated that the situation is very similar for small scale AES examples (the computations needed quite long using Magma and SINGULAR). Moreover, the excellent linear algebra techniques in Magma does not help for linear lexicographical lead rewriting systems.

4.4 Optimization of the Buchberger algorithm with ZDDs

It is possible to use algorithms with ZDDs not only for arithmetic and computing normal forms, but also for much more functionality inside the Buchberger algorithm. In this section we demonstrate several ways, to improve the pair management inside the Buchberger algorithm by set operations with ZDDs.

4.4.1 Classical product criterion

The classical product criterion can be applied immediately, when adding a new generator to the system, as it only depends on the leading terms of the respective polynomials.

In typical Boolean systems (with many variables) this criterion can be applied much more often than in classical areas of Gröbner basis computations. While in the classical case it typically does not apply to most of the pairs, in our applications we have the opposite situation. So the set, where the Buchberger criterion does not apply (the set we are actually interested in), is quite small.

However, usually the number of generators is quite big for our systems. This means that in these examples, the product criterion might apply over a billion times. This shows the extraordinary importance not to compute the criterion for every pair of polynomials. Moreover we observe that due to the large number of variables, classical techniques with short exponent vectors do not work.

Luckily, there exists a solution integrating easily with our ZDD set operations. Using a ZDD we can represent a set of monomials in the same way as a polynomial. For our system of Boolean generators $G = \{g_1, \ldots, g_n\}$, $l_i = \mathrm{lm}(g_i)$ we can represent the set of leading monomials $L = \{l_i | i \in \{1, \ldots, m\}\}$ as ZDD, as well as M, the set of leading terms which are minimal with respect to divisibility. We may assume in the following that each leading term is unique. So we have a one to one correspondence between generators and leading terms. In this way we can associate to a monomial $m \in L$ a polynomial $g_m \in G$. Considering the step in the algorithm when we add a new generator h to the system, we make the following observations: For a monomial $m \in M$ the product criterion applies to the pair (g_m, m) if m does not contain any variables of $\mathrm{lm}(h)$. This can be computed in several ways using ZDD set operations:

1. form the set V of variable of $\mathrm{lm}(h)$. Calculate the multiples of V lying in M by a direct recursive algorithm.

2. take the set M. For each variable x of $\mathrm{lm}(h)$ remove the monomials containing x from M (this operation on ZDDs is usually called subset0).

Note that we actually compute the (small) set of leading monomials where the product criterion, does not apply. The complement forms the majority of pairs and is never taken into account in the complete algorithm which saves a lot of overhead.

While the first solution would be better, as no nodes are created only for intermediate results, at the current time only the second solution is implemented in PolyBoRi.

4.4.2 Minimal sets of pairs

Again, we consider the situation when we add a new polynomial h to our generating system: Using Greuel and Pfister [2002] (theorem 2.5.9) it suffices to add pairs of the form (h, g_m) to our pair set with $m \in M$ (m minimal in L, the set of leading terms), besides the pairs of h with field equations. So, an algorithm for calculating the set of minimal elements M given a set L was needed.

$$M := \{m \in L | \forall l \neq m \in L : l \text{ does not divide } m\}$$

In fact, it turned out that it was quite hard to give a good recursive formulation of this problem which works well with ZDDs. During the development of PolyBoRi, we tried several algorithms (one of variants which turned out to be unsuitable was written by an external developer for the extra-library [Mishchenko, 2003]).

Our solution is to base the algorithm of the minimal elements on algorithm 18: Computing the reduced normal form of a polynomial p against a set of monomials B is equivalent to considering its support $A := \text{supp}(p)$ and restricting it to elements which are not divisible by elements of B:

$$A \mod B := \{a \in A : \nexists b \in B : b \text{ divides } a\} = \text{nf\_mon\_set}(\sum_{a \in A} a, B)$$

Utilizing this normal form computation again as modulo operation between sets, we can give a good formulation of a recursive CUDD-style algorithm for computing minimal elements in a set of monomials. This is done in algorithm 21 and the actual solution is quite short. It shows the full beauty of recursive, functional formulations: Except for some terminal cases, the principle is easy: The minimal elements of a set L with $x = \text{topvar}(L)$ are composed of the minimal elements without x, and the minimal elements of L with x which are not divisible by any (minimal) element without x.

Example 4.4.1. *This example uses the IPython interface of* POLYBORI.

```
In [1]: a=BooleSet([x(1)*x(2),x(2),x(3)])

In [2]: a.minimal_elements()
Out[2]: {{x(2)}, {x(3)}}

In [3]: b=BooleSet([x(1),x(3)])

In [4]: a%b
Out[4]: {{x(2)}}
```

Algorithm 21 Calculating the set of minimal monomials of L

Input: L set of monomials in ZDD representation
Output: result minimal_elements$(L) = \{m \in L | \forall l \neq m \in L : l \text{ does not divide } m\}$
  if $L = \emptyset$ **then**
    return L
  if $1 \in L$ **then**
    return $\{1\}$
  $M_e :=$ minimal_elements(L^E)
  $M_t :=$ minimal_elements$(L^T \mod M_e)$
  return ite(top$(L), M_t, M_e)$

4.4.3 Optimization of the extended product criterion

Another aspect is the application of the extended product criterion: pulling out common factors and then applying the product criterion. In many applications systems contain many polynomials of the form $\prod_j (x_{i_j} + c)$, where $c \in \{0, 1\}$. For these polynomials, we define to sets S_c:

- Such polynomials completely factorizing in the form $\prod_j (x_{i_j} + 1)$.

- Such polynomials completely factorizing in the form $\prod_j (x_{i_j})$ which is equivalent for a Boolean polynomial to be a monomial.

These sets are represented by their leading terms, as these are unique in our polynomial system.

So whenever we add a Boolean polynomial h to our generating system, which is of one those two forms, we consider for the building of pairs only such polynomials, whose leading term does not lie in the corresponding set S_c. We take the difference of sets $M \setminus S_c$, which is yet another very efficient set operation (it does not depend on the number of elements, but only on the ZDD structure). For all pairs (p, h) $(p \in M \setminus S_c)$ the extended product criterion applies as $\frac{p}{\gcd(p,h)}$ and $\frac{h}{\gcd(p,h)}$ do not share any common variables.

Chapter 5

Applications

This chapter presents some applications of Boolean polynomials and Boolean Gröbner bases. Section 5.1 is connected to a problem from computational biology. Moreover, it features methods for computing a reduced lexicographical normal form or the set of lexicographical standard monomials for the ideal of a given set of points without the calculation of a Gröbner basis or a generating system of the ideal.

Section 5.2 is devoted to the application which motivated this thesis: formal verification of integrated circuits. There exist various problems in this domain that can all be modeled using Boolean polynomials.

Since Boolean polynomials are equivalent to Boolean functions, there exists a strong connection to propositional logic. In section 5.3 we provide a method for converting systems of polynomials to the conjunctive normal form which is the native format of the dpll-algorithm [Davis and Putnam, 1960] implemented in SAT-solvers ([Biere et al., 2009]). The proposed converter does not introduce additional variables. This is important for various problems, but in particular useful for algebraic cryptanalysis. For example, Condrat and Kalla experimented with Gröbner bases as preprocessing technique for SAT-solvers. They experienced problems with the large number of additional variables (and clauses). So our conversion algorithm might help to improve the approach given in that article.

Regarding the domain of cryptanalysis we want to refer to a joint work together with Stanislav Bulygin [Bulygin and Brickenstein, 2010] where we broke AES small 64 bits ciphers over two rounds. Before we started this work, this was possible for 8 bits only. Many of the algorithms developed for formal verification also proved to be valuable here, in particular the techniques presented in section 4.3.

5.1 Boolean interpolation

The topic of this section was originally motivated by some problems arising from computational biology, originally stated by Laubenbacher and Stigler [2004]. One important issue in the field of system biology is to detect and model the causal behaviour of the mechanisms in gene regulatory networks [Kitano, 2002].

Laubenbacher and Stigler propose a method for reverse engineering the structure of such networks from experimental data. They have presented an algebraic approach for the generation of Boolean networks in which variables have only two possible states. Such systems can be described by Boolean polynomials.

For this purpose polynomial expressions were generated which match known state combinations (interpolation points) observed during experiments. Since for n variables and k interpolation points, there exist $\binom{n}{k}$ Boolean interpolation polynomials, and since Boolean polynomials are equivalent to Boolean functions, just taking any interpolation polynomial would be completely useless to predict any behaviour. This contrasts with polynomial interpolation over $\mathbb{R}$ where an interpolation polynomial differs in many ways (finitely many zeros, continuity, ...) from a generic function. Therefore, researchers in computational biology try to choose a polynomial with certain mathematical properties and observe how good these properties are suited to predict future experiments. For example interpolation polynomials, that are reduced against the lexicographical Gröbner basis of the vanishing ideal of interpolation points, tend to depend more on the last ring variables than on the first. Since the variables can have different meanings, choosing a good variable ordering and considering reduced polynomials might be a tool to get more meaningful interpolation polynomials.

In the following, a new approach is presented for direct computation of reduced normal forms. The latter can be applied to the reverse engineering problems from computational biology. The algorithm does not need an initial candidate for the interpolation polynomial. Note that the proposed method does not make use of expensive Gröbner basis computations. Indeed, the setup of a polynomial generating system is avoided completely.

Following, special routines for the polynomial interpolation in the Boolean case are presented. Our aim is the generation of multivariate polynomials corresponding to Boolean functions which are partially defined by given pairs of supporting points and function values. As a typical application of numerical mathematics, it was already treated in terms of computational algebra by Mourrain and Ruatta [2002]. More specifically, Armknecht et al. [2006] give a Lagrange-like interpolation scheme which is optimized for the special case of Boolean rings. They have shown that the interpolation problem can be reformulated effectively as a sequence of linear equation systems. This allows to compute a polynomial which fulfills the given partial function definition. It is selected from a vector space whose basis is given as a set of monomials. The latter is extended dynamically until one can verify that it is large enough for generating an interpolation polynomial with the desired properties. By appending monomials to the vector basis according to the given term order one can ensure that a minimal polynomial is generated without the need of Gröbner basis techniques.

In contrast, the methods proposed in this section directly operate on binary decision diagrams corresponding to the input data. Moreover, it is not necessary to compute the vector basis explicitly. The resulting polynomial will automatically contain only standard monomials. In this way, we can also generate results (in the structured case) where the set of standard monomials is too big to be represented directly in a computer. Storing such a basis as a list or vector of monomials fails because of memory limitations. Subsequently, the lexicographically smallest Boolean interpolating polynomial w. r. t. the interpolation data is generated.

In order to utilize the special properties of decision diagrams and Boolean polynomials, this will be formulated in a recursive way. From the computational point of view it is also important to mention that caching of results, which correspond to common subexpressions, can be implemented easily. This improves the overall performance of the algorithm for structured input data which typically occurs in practical examples. In addition, an interpolation-based normal form algorithm will be formulated. It yields further insights about the structure of the lexicographical normal form in the Boolean case.

5.1.1 Interpolation problem

A partial Boolean function $f : \mathbb{Z}_2^n \to \mathbb{Z}_2$ can be defined by two disjoint subsets Z, O of $\mathbb{Z}_2^n$ where $f(o) = 1$ and $f(z) = 0$ for each $z \in Z$, $o \in O$. Given O and Z we can also denote f by b_Z^O. The word *partial* implies that $Z \cup O$ may be a proper subset of $\mathbb{Z}_2^n$. Given two partial Boolean functions $b_{Z_1}^{O_1}, b_{Z_2}^{O_2} : \mathbb{Z}_2^n \to \mathbb{Z}_2$. We define the sum

$$(b_{Z_1}^{O_1} + b_{Z_2}^{O_2}) : \mathbb{Z}_2^n \to \mathbb{Z}_2$$

by the mapping $x \mapsto b_{Z_1}^{O_1}(x) + b_{Z_2}^{O_2}(x)$ for all $x \in (Z_1 \cup O_1) \cap (Z_2 \cup O_2)$. For this partial function definition we search for a Boolean polynomial p where the associated function f_p is a specialization of f, i. e. $f_p(x) = f(x)$ for each $x \in O \cup Z$.

Algorithms for calculating this interpolation can be compared (see section 5.1.5) in performance and compactness of the result. Moreover, it would be worthwhile if the output can be defined by a simple mathematical property.

In the following, we present a method with a very compact result which is lexicographically reduced w. r. t. the vanishing ideal of $O \cup Z$. In this way, it can also be reinterpreted as fast normal form computation against this ideal.

A partial Boolean function can be represented as a pair of two ZDDs Z and O in the following way: A point v in $\mathbb{Z}_2^n$ can be identified in the same way like the monomial x^v with a subset of $\{1, \ldots, n\}$ as $v \leftrightarrow \{i | v_i = 1\}$. In the same way, we can identify a set of points with a subset of the power set of $\{1, \ldots, n\}$ or a ZDD.

In this spirit, the addition of two partial Boolean functions could be expressed as follows in terms of ZDD operations: Let $D = (Z_1 \cup O_1) \cap (Z_2 \cup O_2)$, then

$$(b_{Z_1}^{O_1} + b_{Z_2}^{O_2}) = b_{(Z_1 \cap Z_2) \cup (O_1 \cap O_2)}^{(O_1 \oplus O_2) \cap D}.$$

5.1.2 Encoding

A vector $v = (v_1, \ldots, v_n) \in \mathbb{Z}_2^n$ can be encoded just by the set containing all variables x_i where $v_i \neq 0$. Analogously to polynomials a subset of $\mathbb{Z}_2^n$ can be translated into a subset of the power set of $\{x_1, \ldots, x_n\}$. Using this encoding, we obtain a compact data structure for these kinds of sets which share some common parts of the vectors. The latter may be treated simultaneously by a recursive method. Also, efficient caching of intermediate results for reuse is possible.

This encoding has been defined formally in definition 3.1.20.

5.1.3 Zeros of a Boolean polynomial

In this section, we introduce an algorithm for computing the zeros of a given Boolean polynomial in a (possibly proper) subset of $\mathbb{Z}_2^n$. For illustration, a single Boolean vector $v = (v_1, \ldots, v_n)$ in $\mathbb{Z}_2^n$ is considered. One may define a sequence by setting $p_0 = p$

and subsequently substituting all variables by its corresponding Boolean value in v as $p_i = p_{i-1}|_{x_i=v_i}$. This terminates ultimately in 0 or 1. Hence, it proves whether v is a zero of p or not. For a set of vectors one could do this evaluation element-wise, but this naïve approach is not suitable for large sets. In contrast, using the encoding proposed above, some common parts of the vectors may be treated simultaneously. Again, the resulting subset may be encoded as a decision diagram "on the fly" as algorithm 22 shows.

Algorithm 22 Recursive zeros: zeros(p, S)

Input: Boolean polynomial p, S set of points in $\mathbb{Z}_2^n$ represented as ZDDs
Output: $\text{zeros}(p, S) = \{s \mid s \in S \text{ and } p(s) = 0\}$

```
if p = 0 then
  return S
if (p = 1) or (S = ∅) then
  return ∅
if S = ZDD({∅}) then
  if p has constant part then
    return ∅
  else
    return S
/* As from now p, S are certainly non-constant ZDDs. */
while top(p) < top(S) do
  p = p^E
i = min(top(p), top(S))
p_0 = subset0(p, x_i)
p_1 = subset1(p, x_i)
S_0 = subset0(S, x_i)
S_1 = subset1(S, x_i)
Z_00 = zeros(p_0, S_0)
Z_01 = zeros(p_0, S_1)
Z_11 = zeros(p_1, S_1)
return ite(x_i, S_1\(Z_01 ⊕ Z_11), Z_00)
```

Consequently, algorithm 23 yields the remaining points. These are exactly those points for which a given Boolean polynomial evaluates to one.

Algorithm 23 Ones: ones(p, S)

Input: Boolean polynomial p, S set of points in $\mathbb{Z}_2^n$
Output: $\text{ones}(p, S) = \{s \mid s \in S \text{ and } p(s) = 1\}$

```
return S\zeros(p, S)
```

5.1.4 Normal forms against a variety

Normal forms are very hard to compute, in particular if calculations of Gröbner bases are involved. So, it is a quite natural question whether the latter may be avoided by introducing specialized methods for calculating normal forms. The presented method works even without a generating system of the ideal, i. e. in case it is just given in form of its variety.

Since ideals in the ring of Boolean polynomials are in one-to-one correspondence to subset of $\mathbb{Z}_2^n$, this problem is quite general. Moreover, the Gröbner basis of the ideal can be a much bigger object than the corresponding variety.

Definition 5.1.1. *Let $>$ be an arbitrary monomial ordering, then we can extend $>$ lexicographically to the set of Boolean polynomials (we make use of the fact that all non-zero coefficients are one) by setting $p \geq q$ if and only if one of the following conditions holds:*

- $q = 0$,
- $p \neq 0$, $q \neq 0$ *and* $\operatorname{lm}(p) > \operatorname{lm}(q)$,
- $p \neq 0$, $q \neq 0$, $\operatorname{lm}(p) = \operatorname{lm}(q)$ *and* $\operatorname{tail}(p) \geq \operatorname{tail}(q)$.

Lemma 5.1.2. *Let $I \supset \langle x_1^2 + x_1, \ldots, x_n^2 + x_n \rangle$ be an ideal with $I \subset \mathbb{Z}_2[x_1, \ldots, x_n]$, p a Boolean polynomial, and let G be a Gröbner basis of I. Then the following two statements are equivalent:*

1. *p is the lexicographically smallest Boolean polynomial in $p + I$.*
2. $\operatorname{REDNF}(p, G) = p$ *(i. e. p is reduced).*

Proof. A polynomial is reduced if and only if all its terms are reduced (standard monomials).

Let us assume that a term t of p is not reduced: Then there exists an $f \in G$, s. th. t can be rewritten by f modulo I. Applying the field equations the result is again a Boolean polynomial. In this way, we have constructed a smaller Boolean polynomial in $p + I$.

On the other hand, let p be reduced. Assume, there exists a smaller polynomial q in the same residue class. We may assume that $\operatorname{lm}(p) \neq \operatorname{lm}(q)$. Then we know from general Gröbner basis theory that $\operatorname{REDNF}(p, G) = \operatorname{REDNF}(q, G)$. Since p is reduced, it follows $p = \operatorname{REDNF}(q, G)$ and $p - q$ has a standard representation (each summand has a leading term smaller or equal to $\operatorname{lm}(p) = \operatorname{lm}(p - q)$).

Hence, it follows that one of the summands in this representation has $\operatorname{lm}(p)$ as leading term and p is not reduced. □

In order to show the basic principles of interpolating with Boolean polynomials, we provide a simple interpolation procedure given in algorithm 24. If two sets Z and O of vectors are given, we can reduce the interpolation problem to the case where the i-th index of all vectors is 0. For this purpose let Z_0, O_0 denote the corresponding sets of those vectors whose i-th entry is zero (subset-0 points). Analogously, for given $S \in \{Z, O\}$ let the set $S_1 = \{(v_1, \ldots, v_{i-1}, 0, v_{i+1}, \ldots v_n) \mid v \in S \text{ and } v_i = 1\}$ denote the subset-1 points.

First, we calculate an interpolation h_e for $Z_0 \cup 0_0$. In the second step, we may continue with an interpolation h of the subset-1 points in $Z_1 \cup O_1$. One could combine

Algorithm 24 interpolate_simple(b_Z^O)

Input: b_Z^O a partial function definition.
Output: interpolate_simple(b_Z^O) $= p$ with $f_p = b_Z^O$ on $Z \cup O$

```
if Z = ∅ then
    return 1
if O = ∅ then
    return 0
i = min(top(O), top(Z))
Z_1 = subset1(Z, x_i)
Z_0 = subset0(Z, x_i)
O_1 = subset1(O, x_i)
O_0 = subset0(O, x_i)
h_e = interpolate_simple(b_{Z_0}^{O_0})
h_t = interpolate_simple(b_{Z_1}^{O_1}) + h_e
return x_i · h_t + h_e
```

both recursively generated results to $x_i \cdot h + h_e$, but this yields wrong results for vectors in $O_1 \cap O_0$. This can easily be fixed by cancelling those terms already treated by h_e. In this case, it can be done just by adding h_e to h because the terms with x_i do not influence the behaviour on $Z_0 \cup O_0$. For this reason h_e is computed first.

On the other hand, if we really want to have a minimal interpolation polynomial, then we should not adjust the bigger terms (those with x_i: $x_i \cdot h_t$) to the smaller terms (those without x_i), but do the opposite. So we would like to compute h_t first. The problem is if we want to calculate the then-branch of the interpolation polynomial h_t, then we have to know the value which h_e takes on the interpolation points specified for h_t. Moreover, we should not specify more interpolation points than needed for h_t as for every point in $Z_1 \cup O_1$ which does not lie in $Z_0 \cup O_0$ (subset-0 points) the value of the function can be adjusted by h_e. Luckily the behaviour of h_e on $(Z_1 \cup O_1) \cap (Z_0 \cup O_0)$ is predictable and this gives exactly the set of interpolation points which we will specify for h_t. This approach is incorporated in algorithm 25.

Example 5.1.3. *We want to compute a minimal interpolation polynomial for b_Z^O using algorithm 25 where*

$$Z = \{(0,0,0),(1,1,0),(1,0,1)\}\,,\; O = \{(1,0,0),(0,1,0),(0,1,1)\}\,.$$

In order to reduce the problem, we compute the cofactors w. r. t. the first variable and skip the first component which is always zero. This yields

$$Z_1 = \{(1,0),(0,1)\},\; O_1 = \{(0,0)\},\; Z_0 = \{(0,0)\},\; O_0 = \{(1,1),(1,0)\}\,.$$

Hence $f + g = b_\emptyset^{\{(1,0),(0,0)\}}$ which implies $h_t = 1$. Here, we get a first impression of the reason why the algorithm actually returns the minimal interpolation polynomial: Since h_t is finally multiplied by x_0, it is the lexicographically biggest part of our computation. We specify a minimal set of interpolation points. In this case, it is even possible to fulfil this partial function definition by the constant polynomial 1. For

$$w = b_{\{(0,0)\}}^{\{(1,1),(1,0),(0,1)\}}$$

Algorithm 25 interpolate_smallest_lex(b_Z^O)

Input: b_Z^O a partial function definition.
Output: interpolate_smallest_lex(b_Z^O), smallest Boolean polynomial p w. r. t. lex. with $f_p = b_Z^O$ on $Z \cup O$

```
if Z = ∅ then
  return 1
if O = ∅ then
  return 0
i = min(top(O), top(Z))
Z_1 = subset1(Z, x_i)
Z_0 = subset0(Z, x_i)
O_1 = subset1(O, x_i)
O_0 = subset0(O, x_i)
/* C forms the set of conflict points between 0 and 1-subsets */
C = (Z_1 ∪ O_1) ∩ (Z_0 ∪ O_0)
f = b_{Z_1}^{O_1}
g = b_{Z_0}^{O_0}
h_t = interpolate_smallest_lex(f + g) /* f+g is only defined on C */
F = ones(h_t, ((Z_1 ∪ O_1)\C))
/* non-conflict subset1 terms affected by else branch */
w = b_{((Z_1\C)⊕F)∪Z_0}^{((O_1\C)⊕F)∪O_0}
h_e = interpolate_smallest_lex(w)
return x_i · h_t + h_e
```

we recursively obtain $h_e = \text{interpolate\_smallest\_lex}(w)$, *which equals* $x_1 \cdot x_2 + x_1 + x_2$. *This gives us the result:*

$$\begin{aligned}\text{interpolate\_smallest\_lex}(b_Z^O) &= x_0 \cdot h_t + h_e \\ &= x_0 + x_1 \cdot x_2 + x_1 + x_2\end{aligned}$$

The following results show that the algorithm computes the desired interpolation.

Lemma 5.1.4. *Let S be a set of non-constant Boolean polynomials, v the smallest top variable occurring in S (biggest index: $\max\{\text{top}(f)|f \in S\}$), $S \ni p = v \cdot p_1 + p_0$ (v not occurring in p_1, p_0). Then p is the lexicographically smallest polynomial w. r. t. lexicographical monomial ordering if and only if p_1 is the minimal cofactor in S and for all $g \neq p$ with $g = v \cdot p_1 + g_0$ holds: $g_0 > p_0$.*

Proof. This lemma is a consequence of the fact that if $p = v \cdot p_1 + p_0$ with v as top variable, all terms in which v occurs are bigger than those without v. So these are considered first in lexicographical comparison of Boolean polynomials. □

Theorem 5.1.5. *Algorithm 25 returns an interpolation polynomial which is the lexicographically smallest polynomial w. r. t. lexicographical monomial ordering under all polynomials interpolating the same function on P.*

Proof. The algorithm checks first the cases where Z or O are empty. So both are non empty. Since they are disjoint, we know that at least one of them contains a nonzero vector. Hence, we can assume that all recursive calls deliver a polynomial with the desired property and that i is contained in $\{1, \ldots, n\}$.

It can be checked easily that the result provides an interpolation of the specified function by doing case considerations.

To see that it is minimal, we use the previous lemma for minimality: First, we have to check that h_t is chosen minimal. By our assumptions the recursion returns a minimal polynomial under the specified properties. Thus, we know that it is minimal under all possible choices for the then-branch h_t of our interpolation polynomial, fulfil the specified condition (which means that the recursive call chooses the minimal polynomial out of the full set of candidates). We just have to check that all conditions are necessary for the then-branch of an interpolation polynomial.

In fact, we only prescribe the behaviour of h_t on all conflicting points, such originally prescribed points, which occur in both combinations: those, with index i set to 0 on the one hand (denoted by C_0), and index i set to 1 on the other (C_1). As the set C in the algorithm is already projected to 0 in the i-th component, we can denote this by

$$C_1 = \{c_{x_i=1} | c \in C\} \text{ and } C_0 = C\,.$$

For these points we have the following situation:

$$\forall c \in C_0 : h_e(c) = h_e(c_{x_i=1}), c_{x_i=1} \in C_1, (x_i \cdot h_t)(c) = 0\,.$$

No matter what h_t we choose, we get for every suitable h_e

$$h_e(c) = 1 \Leftrightarrow c \in O_0 \text{ for all } c \in C_0.$$

We conclude that for all $c \in C_1$ the equality $h_t(c) = 1$ holds if and only if

$$(c_{x_i=0} \in O_1 \text{ and } c_{x_i=0} \in Z_0) \text{ or } (c_{x_i=0} \in O_0 \text{ and } c_{x_i=0} \in Z_1)\,.$$

These conditions on h_t are equivalent to the partial Boolean function definition $f+g$. Thus the specification $f+g$ for our then-branch contains only necessary interpolation points and recursively we get a minimal interpolation polynomial for this part. By this we fulfilled the first condition of our lemma and calculated a minimal h_t under all interpolation polynomials. Note that if we branch at an index i, which is not the biggest possible for an interpolation, h_t will be just zero. Therefore, this is also compatible with the lemma.

Having the correct h_t, we have to check, that under all Boolean interpolation polynomials of the form $x_i \cdot h_t + h_e$, we choose a minimal h_e. The behaviour as function of h_e is uniquely determined on $Z \cup O$ by h_t and the prescribed interpolation values, and this is just what our recursive call for calculating h_e does: passing these (possibly adjusted) values. Since all candidates for h_e need to fulfil this condition, we really get the full sets of candidates and by recursion a minimal else-branch. □

The following theorem combines the well-known Propositions 4 and 8 from [Cox et al., 2007, Chapter 5, §3].

Theorem 5.1.6. *For each monomial ordering $>$ and each set of points P the set of standard monomials S w. r. t. $\mathrm{I}(P)$ in $\mathbb{Z}_2[x_1, \dots, x_n]$ has the same cardinality as P:*

$$|S| = |P| .$$

Hence, we have an obvious upper bound for the number of terms of reduced normal forms.

Corollary 5.1.7. *Let $f \in \mathbb{Z}_2[x_1, \dots, x_n]$ be a Boolean polynomial, P be a set of points in $\mathbb{Z}_2^n$. Then a reduced normal form $g = \mathrm{REDNF}(f, \mathrm{I}(P))$ has at most $|P|$ terms.*

Algorithm 22 and algorithm 25 can be combined to compute a reduced normal form without using Gröbner bases.

Algorithm 26 Reduced lexicographical normal form against variety
nf_by_interpolate(f, P)

Input: Boolean polynomial f, P set of points in $\mathbb{Z}_2^n$.
Output: nf_by_interpolate$(f, P) = \mathrm{NF}(f, \mathrm{I}(P))$
  $Z = \mathrm{zeros}(f, P)$
  return interpolate_smallest_lex$(b_Z^{P \setminus Z})$

Theorem 5.1.8. *Algorithm 26 calculates a lexicographical normal form of f w. r. t. a Gröbner basis of the ideal*

$$\mathrm{I}(P) = \{g \in \mathbb{Z}_2[x_1, \dots, x_n] \mid g(p) = 0 \text{ for all } p \in P\} .$$

Proof. First note: A reduced normal form against a Gröbner basis G does not depend on the particular choice of G, but only on the ideal generated by G. Moreover, since Boolean polynomials are in one-to-one correspondence to Boolean functions, each Boolean polynomial representing the same function on P lies in the same residue class modulo $\mathrm{I}(P)$. A normal form $g = \mathrm{NF}(f, G)$ of a polynomial against a Gröbner basis G can be characterized to be the unique reduced polynomial, since $g = \mathrm{NF}(g, G)$ where $g + \langle G \rangle = f + \langle G \rangle$. So we have to show for given f, and P:

1. nf_by_interpolate(f, P) represents the same function restricted to P as f.

2. nf_by_interpolate(f, P) is reduced.

The first claim holds by the correctness as we have already proven, that nf_by_interpolate returns indeed an interpolating function. The reducedness follows from Lemma 5.1.2. □

Next, algorithm 27 computes the standard monomials of an ideal of a variety which is given as a set of points.

The minimal elements of the remaining Boolean monomials are exactly the leading monomials of the minimal Gröbner basis. Hence, we can formulate algorithm 28 which can be used to determine the size of the basis without explicitly computing it.

Algorithm 27 Standard monomials of $\mathrm{I}(P)$: standard_monomials_variety(P)

Input: P a set of points in $\mathbb{Z}_2^n$.
Output: $S = \{t \mid \exists \text{ reduced } p \in \mathrm{I}(P) : t \text{ term of } p\}$
  $S = \emptyset$
  while $|P| \neq |S|$ **do**
    $Z = \text{random\_subset}(P)$
    $p = \text{interpolate\_smallest\_lex}(b_Z^{P \setminus Z})$
    $S = S \cup \text{supp}(p)$
    $S = \{t \text{ term } |\exists s \in S : t \text{ divides } s\}$
  return S

Algorithm 28 Leading monomials of a minimal Gröbner basis of $\mathrm{I}(P)$

Input: P set of points in $\mathbb{Z}_2^n$.
Output: leading_monomials_variety$(P) = \mathrm{L}(\mathrm{I}(P))$
  $T = \{t \text{ Boolean term in } \mathbb{Z}_2[x_1, \ldots, x_n]\}$
  $R = T \setminus \text{standard\_monomials\_variety}(P)$
  return minimal_elements(R)

Theorem 5.1.9. *Assume that the probability distribution of the choice of random subsets Z are independent equipartitions for each point $p \in P$ and form independent experiments. Then the probability p for the main loop of Algorithm 27 passing more than k iterations is less or equal to*

$$1 - (1 - \frac{1}{2^k})^{|P|}.$$

Proof. For the analysis for the algorithm, we can ignore the step including term which divide already found terms as this only reduces the number of needed iterations. For a random subset $Z \subset P$ the probability that a standard monomial t occurs in the interpolation is $\frac{1}{2}$. The probability that a term t does not occur in k experiments is $\frac{1}{2^k}$. Since the probability distribution of independent equipartitions on P result in independence on the set of standard monomials, the probability that all $|P|$ standard monomial occur in k experiments is $(1 - \frac{1}{2^k})^{|P|}$. □

All presented set operations – except the random_subset procedure – have been implemented in a decision diagram-style recursive approach. In particular, there is no problem to represent the set of terms T with 2^n elements in it as the number of ZDD-nodes is just n (in this case).

Algorithm 29 Reduced lexicographical Gröbner basis of $\mathrm{I}(P)$

Input: P set of points in $\mathbb{Z}_2^n$.
  return $\{t + \text{nf\_by\_interpolate}(t, P) \mid t \in \mathrm{L}(\mathrm{I}(P))\}$

Algorithm 29 shows that it is even possible to compute the complete Gröbner basis of $\mathrm{I}(P)$ just having this normal form algorithm. However – the actual advantage of this section is the computation of normal forms without generating the Gröbner basis.

5.1.5 Practical Experiments

In this section, we consider randomly generated examples and show that we can still compute lexicographical normal forms where the computation of the full Gröbner basis seems to be practically unfeasible. These generating systems are usually quite big. In fact for Table 5.1 with our experiments, the size of the Gröbner basis was computed by algorithm 28 which returns the leading ideal. This is much cheaper than computing the Gröbner basis itself (at least using algorithm 29 as you can see in the following). If you consider the size of the Gröbner basis s, the number of points $|P|$, the cost of a single normal form N. Then the cost of computing the leading ideal is approximately $\log(|P|) \cdot N$ and for the Gröbner basis $s \cdot N$ (if you like, you can assume that s has the same dimension as $|P|$). This is of course no exact complexity analysis, but gives impression of the computation problem.

For computing the normal form against the variety, neither the Gröbner basis is needed, nor the leading ideal, but just a single call of nf_by_interpolate. This is much easier, in particular because the Gröbner basis in the biggest example in this listing has about 600 000 elements. Due to the random nature of our examples we expect the tail of each basis element quite dense in the set of standard monomials which has size $|P|$, 500 000 in this example.

Our timings have been done on an AMD Dual Opteron 2.2 GHz (we have used only one CPU) with 16 GB RAM running Linux. We used random sets of points and random partial functions in 100 variables. Note that random data is supposed to be the worst case scenario for PolyBoRi as using caching techniques our algorithms work better on structured polynomials.

We were not able to run interpolate_simple on the bigger example since the memory consumption was too high. On the other hand, this algorithm shows nice performance in the small examples (mainly as the expensive call to zeros is missing). In this sense, these results confirm that our algorithm is able to compute normal forms without Gröbner basis up to a magnitude in which it seems impossible to compute the Gröbner basis on nowadays machines in reasonable time.

5.1.5.1 Structured Examples

The results illustrated in Table 5.1 clearly indicate that the proposed procedure can be used for handling large sets of points. But the PolyBoRi framework cannot show its full potential here because the example was generated from random data. Hence, the points were chosen rather generically. In contrast, PolyBoRi can deal even better with structured data as present in real-world examples. In order to obtain such a kind of benchmark in a comprehensible and scalable way, we consider a series of examples derived from J. Conway's *Game of Life*, see Gardner [1971]. It has already been used by Gerdt and Zinin [2008] to compare the performance of their Boolean Gröbner basis implementation with those of Singular and PolyBoRi. Each example `life(i)` is defined by a polynomial in $i+1$ variables $x_0, \ldots, x_i$ of the form

$$p_i = x_i + x_{i-1}(\sigma_{i-2} + \sigma_{i-3} + \sigma_3 + \sigma_2) + \sigma_{i-2} + \sigma_3\,, \tag{5.1}$$

where σ_k is the symmetric polynomial of degree k in $x_0, \ldots, x_{i-2}$. The zero set of p_i is used as the set of supporting points for a partial function definition. In order to have a

#	smallest lex.		interpolate_simple		#
points	time	length	time	length	basis
100	$0.01s$	42	$0.00s$	12771	287
500	$0.06s$	249	$0.01s$	$\approx 2.9 \cdot 10^{10}$	1943
1000	$0.29s$	508	$0.01s$	$\approx 8.1 \cdot 10^{13}$	3393
5000	$5.53s$	2552	$1.47s$	$\approx 4.5 \cdot 10^{23}$	10319
10000	$19.78s$	5020	$37.18s$	$\approx 1.6 \cdot 10^{26}$	17868
50000	$250.95s$	25012			82929
100000	$897.85s$	50093			162024
200000	$3488.61s$	99868			296697
500000	$20336.02s$	249675			636542

Table 5.1: Random interpolation problems in 100 variables: We measured the size of the number of terms for each interpolation polynomial and determined the number of elements in a minimal Boolean Gröbner basis of $\mathrm{I}(\mathrm{V}(O \cup Z))$ without calculating the actual polynomials.

Example	Variables	Prescribed Ones		Prescribed Zeros		Result			
		Points	Nodes	Points	Nodes	Terms	Nodes	Degree	Time
`life(10)`	11	660	63	364	50	268	58	9	$0.00s$
`life(50)`	51	$7.03 \cdot 10^{14}$	503	$4.22 \cdot 10^{14}$	410	39348	458	49	$0.01s$
`life(100)`	101	$7.92 \cdot 10^{29}$	1053	$4.75 \cdot 10^{29}$	869	323698	958	99	$0.01s$
`life(500)`	501	$2.04 \cdot 10^{150}$	5453	$1.22 \cdot 10^{150}$	4469	41418498	4958	499	$0.09s$
`life(1000)`	1001	$6.69 \cdot 10^{300}$	10953	$4.01 \cdot 10^{300}$	8969	332336998	9958	999	$0.22s$
`life(2500)`	2501	$> 10^{308}$	27453	$> 10^{308}$	22469	5202092498	24958	2499	$0.59s$
`adder(16)`	49	2155872256	204	2139095040	204	131054	75	17	$0.00s$
`adder(32)`	97	$9.22 \cdot 10^{18}$	428	$9.22 \cdot 10^{18}$	428	8589934558	155	33	$0.01s$
`adder(64)`	193	$1.70 \cdot 10^{38}$	876	$1.70 \cdot 10^{38}$	876	$3.68 \cdot 10^{19}$	315	65	$0.01s$
`adder(128)`	385	$5.78 \cdot 10^{76}$	1772	$5.78 \cdot 10^{76}$	1772	$6.80 \cdot 10^{38}$	635	129	$0.02s$
`adder(256)`	769	$6.70 \cdot 10^{153}$	3564	$6.70 \cdot 10^{153}$	3564	$2.31 \cdot 10^{77}$	1275	257	$0.03s$
`adder(512)`	1537	$8.98 \cdot 10^{307}$	7148	$8.98 \cdot 10^{307}$	7148	$2.68 \cdot 10^{154}$	2555	513	$0.05s$

Table 5.2: Structured examples.

nontrivial partition into points which map to zero and one, respectively, we decided to use parity decomposition here, i. e. points with an even number of nonzero entries mark the set of zeros for the interpolation problem while the remaining points define the ones. Of course, a parity decomposition of any set may be interpolated by the polynomial $\sum_0^i x_i$, but the latter may not have the desired minimality property. As it can be seen from the number of terms, in our examples the minimal interpolation polynomial looks much different. The first part of Table 5.2 summarises the complexity of input data and results as well as the computation time for some of these examples.

In addition, `adder(n)` denotes a problem based on a bit-level implementation of an n-bit adder which was modelled in terms of Boolean polynomials by Brickenstein and Dreyer [2009a]. Let $a = (a_0, \ldots, a_{n-1})$ and $b = (b_0, \ldots, b_{n-1})$ be the variable vectors of the inputs which are given as n-bit words. The output variables corresponding to an n-bit word and one carry bit are denoted by $c = (c_0, \ldots, c_n)$. Following, we select those points as supporting points for which the n-bit addition $a + b = c$ holds. This can be described by $c_i = a_i + b_i + carry_{i-1}$ for $i = 0, \ldots, n-1$, and $c_n = carry_n$. Each $carry_i$ denotes a Boolean polynomial which is defined recursively by $carry_{-1} = 0$

and $carry_k = a_k \cdot (b_k + carry_{k-1}) + b_k \cdot carry_{k-1}$. Again, the above decomposition is used to define the sets of ones and zeros. The results are also shown in Table 5.2.

One can conclude that the proposed procedures can easily handle problems arising from very large supporting sets, in particular if those are structured. The (almost instantaneous) result is computed in such a way that intermediate results will not blow-up extensively. Also, PolyBoRi's sophisticated data structures allow to store the result, even though it has a large number of terms. This is due to the fact that the number of decision diagram nodes stays rather small in contrast to the number of terms.

5.2 Formal verification

We designed the PolyBoRi framework for computing Gröbner bases of systems derived from problems in verification of models for integrated circuits. Often, these systems feature a large number of variables and polynomial equations (several thousands and more). So it was a primary design goal to achieve a compact representation for this data. On the other hand, despite better properties of the case of Boolean polynomials [Tran and Vardi, 2007], it is known that there exist systems with an even smaller number of variables (e. g. 50 variables), whose Gröbner bases are hard to compute [Faugère, 2003].

5.2.1 Integrated circuits topology

We start applying computational algebra to combinatorial networks, i. e. networks containing no memory of previous states (flip-flops). Combinatorial networks consist of signals which can be modeled as Boolean variables and logic gates defining a functional relation between these variables.

Each logic gate takes several input signals and transforms them into exactly one output signal. For example an OR-gate transforms two signals y, z into a third signal x by $x := y \vee z$. So the variable x depends on y and z. In some sense x is used to replace larger expressions. Using the algebraic mapping (see definition 2.1.8) from propositional logic this dependency is equivalent to the Boolean polynomial $f := x - y \cdot z$. We would like to have a monomial ordering where the application of the reduced normal form of a Boolean polynomial g against the system consisting of f and the field polynomials replaces all occurrences of x by $y \cdot z$. So, we need to have $\mathrm{lm}(f) = x$ (or $x > y \cdot z$). For a lexicographical ordering this is equivalent to $x > y$ and $x > z$. In this way, each logic gate gives a condition for the monomial ordering. We call a monomial ordering **topological**, if it matches these conditions. A variable ordering is called topological, if the lexicographical ordering with respect to this variable ordering is topological. A topological ordering has not necessarily to be a lexicographical or elimination ordering. It can even be realized by a weighted degree ordering.

Usually there exist several topological variable orderings. Since every combinatorial network forms a directed **acyclic** graph, there always exists such a variable ordering. It can be determined by algorithm 30.

Theorem 5.2.1. *Algorithm 30 is correct and terminates*

Proof. The only critical point of the algorithm is the validity of the "choose" statement. Initially we can choose just any signal and then follow the flow of signal until we reach

Algorithm 30 Calculating topological ordering on G,V

Input: G set of logic gates, V set of variables
Output: return tuple of variables in V topologically ordered w. r. t. G
  if $V = \emptyset$ **then**
    return V
  choose $v \in V$ where v is not an input of g for all $g \in G$
  if $\exists g \in G$: v is the output of g **then**
    $G = G \setminus \{g\}$
  return $(v) + \text{topological\_ordering}(G, V \setminus \{v\})$ /* concatenation of tuples */

some end point (which is not an input to a gate). This is a valid candidate for v. Here we use the fact that the combinatorial network forms a directed acyclic graph, i. e. it contains no loops. It is easy to see that taking v as biggest variable does not contradict any topology condition from the gates: If there exists a g with output v, then making v the biggest variable of all satisfies the condition that v is bigger than the gates input variables variables. The other gates/equations/polynomials do not involve v, so correctness can be satisfied recursively. Since in each recursion step $|V|$ decreases, there can only be finitely many recursion steps and hence the algorithm terminates. □

Corollary 5.2.2. *For each combinatorial network there exists a topological variable ordering.*

Remark 5.2.3. *Topological orderings form the precondition for formal verification with computational algebra. For most other orderings like degree orderings or lexicographical orderings with a generic ordering of variables, it takes already seconds to do an equivalence check on a 4-bit multiplier and 6-bits turn out to be unfeasible. Using advanced techniques and (even optimized) topological orderings, we were able to verify a* 16×16*-bit multiplier in 24 seconds, as it is shown in Table 5.3. This is remarkable, as the the case of multipliers is very difficult. As stated by our project partners [Wedler, 2007], using the industrial standard SAT-solvers, problems larger than* 10×10 *can hardly be handled. We want to remark at this point that we indeed provide some extra information to the Gröbner basis/normal form algorithm in form of a topological ordering.*

The results on combinatorial networks can be generalized to sequential circuits (integrated circuits containing state information) in the following way: Sequential circuits problems can be mapped to propositional logic by using a bound t for the time steps (bounded model checking). A signal x is identified with t distinct Boolean variables. In this way we can construct a directed, acyclic graph and apply the techniques presented for combinatorial circuit in this section.

5.2.2 Optimized topological ordering

Algorithm 30 contains many choices. In general the topological ordering is not unique. Moreover, the choices can influence the strategy of the algorithms. Hence, the question arises, whether some topological orderings have better properties than others.

We recursively define the rank of a signal:

$$\text{rank}(v) = \max(\{\text{rank}(w) + 1 | \exists \text{ gate } g : v \text{ is input of } g, w \text{ is output of } g\} \cup \{0\}).$$

There exists a wide variety of rank concepts, e. g. in [Gansner et al., 2006] , [Ilsen et al., 2008]. Sometimes also the term "level" is used synonymously.

This reflects the usual behaviour that term substitution blows up with every substitution step and we want to keep the intermediate results during the normal form computation as slim as possible.

It seems preferable to sort the variables in ascending order with respect to their rank.

Theorem 5.2.4. *Let $>$ be an ordering on the variables which is compatible with the* rank-*function:*

$$\forall i, j : \mathrm{rank}(x_i) > \mathrm{rank}(x_j) \Rightarrow x_i < x_j.$$

Then $>$ is a topological variable ordering.

Proof. If a variable or signal v forms an input value on the gate determining w as output-function, then we have

$$\mathrm{rank}(v) \geq \mathrm{rank}(w) + 1 \Rightarrow \mathrm{rank}(v) > \mathrm{rank}(w) \Rightarrow v < w,$$

which proves the claim by the definition of the topological variable ordering. □

Theorem 5.2.5. *For each i from 0 to $d := max(\{\mathrm{rank}(x_i) | i \in \{1, \ldots, n\}\})$ define*

$$X_i = \{x_j | \, \mathrm{rank}(x_j) = i\}.$$

We define the following orderings:

- *On each set of variables X_i we define $>_i$ an monomial ordering for $\mathbb{Z}_2[X_i]$. On $\mathbb{Z}_2[x_1, \ldots, x_n]$ we define $>_p$ to be the product ordering of $>_0, >_1, \ldots, >_d$.*
- *Let $>_h$ be an arbitrary monomial ordering on $\mathbb{Z}_2[x_1, \ldots, x_n]$. Then define $>_q$ to the product of $\deg_{X_0}, \ldots, \deg_{X_d}, >_h$. We repeat that the function $\deg_{X_i}$ defines the total degree in the variables contained in X_i.*

Then the orderings $>_p$, $>_q$ are topological.

Proof. Let $> \in \{>_q, >_p\}$, g a gate, x_o the output of g, and $x_{k_1}, \ldots, x_{k_r}$ the inputs of g (for some r). Then for all $j \in \{1, \ldots, r\}$ we have that

$$\mathrm{rank}(x_{k_j}) \geq \mathrm{rank}(x_o) + 1 > \mathrm{rank}(x_o).$$

Since $>$ has the elimination property for $X_{\mathrm{rank}(x_o)}$ restricted to $\mathbb{Z}_2[X_i | i \geq \mathrm{rank}(x_o)]$, x_o is bigger than every monomial and in $x_{k_1}, \ldots, x_{k_n}$. So the topological condition for g is fulfilled. □

Next, we give an algorithmic description, how to compute the rank for all signals in the following Python code.

```
def rank(data):
    parents=dict()
    res=dict()
    for p in data:
```

```
        out=output_signal(p)
        parents.setdefault(out, [])
        for v in input_signals(p):
            parents.setdefault(v, []).append(out)
    def do_rank(v):
        if v in res:
            return res[v]
        my_res = res[v] = max([do_rank(p)+1
            for p in parents[v]]
            + [0])
        return my_res
    for v in parents.keys():
        do_rank(v)
    return res
```

5.2.2.1 Circuit polynomials

We denote the set of polynomial conditions derived from the logic gates of an integrated circuit C by E_C. Furthermore, let I_C be the ideal generated by E_C and the field polynomials. In this way $\mathrm{V}(I_C)$ denotes the set of all possible configurations of signals: all values for signals which are compatible with the logic gates of the circuit. Whenever we talk about the circuit, we may assume that all polynomial equations in E_C are fulfilled.

5.2.3 Formal equivalence checking

Formal equivalence checking is a quite important part of the verification process. Given two integrated circuits, one has to prove that they yield the same output on the same input signals [Krautz et al., 2008].

This technique does not prove, whether the circuit actually does, what it is supposed to do. It just checks that the output signals form the same Boolean function. Equivalence checking usually takes place after small design steps: For instance, after applying an optimization to the integrated circuit it is necessary to check that the optimization did not alter the behaviour of the integrated circuits. Usually these modifications are quite subtle, s. th. major parts of the original design structure are preserved. In this way typical tools for equivalence checking use exactly these structural similarity to make this comparison very fast. On the other hand, usually these tools are not suited for comparing two circuits which were constructed independently.

Algebraic methods can combine the advantages of the two methods to some extent. The circuits are usually given only with identified input variables. Of course, if we have polynomials $x + f$ and $y + f$ with leading monomials x and y, then we can replace all occurrences of x in the equations by y. This reduces the number of variables (using the similar structure) and makes further optimizations possible. In PolyBoRi, there exists an option `eliminate_identical_variables` for applying this technique.

Usually equivalence checking is combined with other verification methods which really verify the behaviour or properties of the respective circuits. This can consume a lot of time. Usually having proven these properties once, designers only need to check the equivalence of their work with the previous design. It can be done in a quite decent time by these specialized tools.

Our experiments indicate that we are able to prove the equivalence of two multipliers in a time that is better than exponential in the number of input bits of the integrated circuit. Note that the total number of variables is still much higher. Handling so many variables was only possible using an optimized topological monomial ordering and the llnf*-algorithm for normal form computation (section 4.3). The time for optimizing the monomial ordering is included in the results shown in Table 5.3. All times are in seconds. On the one hand we measure the time using only our normal form algorithms for each example. On the other hand, we combine it with structural optimization which identifies identical variables in both integrated circuits.

We are very aware that FDDs for multipliers have an exponential number of nodes [Becker et al., 1995]. Our experiments, however, do not indicate exponential behaviour. Indeed, the result of the normal form computation is the zero polynomial, so we do not compute these exponential size FDDs. However, they can appear in intermediate results. Using the optimized topological ordering in connection with the llnf* algorithm seems to lead to the following desired effect: At the same time, both circuits are expanded at similar signals and the terms of the polynomial can be extinguished early in the computation.

Example	time	time (structured)	num. of input var.	total num. of var.
multiplier 8x8	0.09	0.07	16	204
multiplier 10x10	0.32	0.14	20	314
multiplier 16x16	76.05	23.65	32	768

Table 5.3: Equivalence check of multipliers using llnf* in an optimized topological ordering.

5.2.4 Property checking

Given a design it is desired to know, if a certain set of assertions is always satisfied. This assertions are usually called properties and the process of verifying them is known as "property checking". For a more detailed introduction we would like to refer to [Bérard et al., 1999]. In the following we will restrict ourselves to the case of bounded property checking (see 5.2.1) that can be handled by propositional logic.

Algebraically the design/circuit C is represented by the ideal I_C. The most basic way to represent a property is a Boolean function or a Boolean polynomial p.

Definition 5.2.6. *A property p* ***holds*** *on a point v, if and only if $p(v) = 0$. It is said that p holds under a condition c, if and only if $p(v) = 0$ for all $v \in \mathrm{V}(c)$.*

Theorem 5.2.7. *A circuit C fulfills the property p if and only if* $\mathrm{REDNF}(p, I_C) = 0$.

Proof. By Hilberts Nullstellensatz

$$p \in \sqrt{I_C} = I_C \text{ if and only } p(b_1, \ldots, b_n) = 0 \text{ for all } (b_1, \ldots, b_n) \in \mathrm{V}(I_C)$$

holds. □

Remark 5.2.8. *Using a topological monomial ordering the polynomial conditions E_C from section 5.2.1 form a Boolean Gröbner basis of I_C. So we only need to compute one normal form to check p.*

Usually properties are given much more structured, than by a single Boolean expression/polynomial. They are formulated in languages designed for logic like Verilog [IEEE Computer Society, 1996].

Often they are called theorems. Like theorems in mathematics you can structure them in an if-condition i and a claim c in the then-part.

Definition 5.2.9. *The theorem or property* ***"if** i **then** c**"** *is defined to be* $(1+i)\star c$. *A property "if i then c" holds for a circuit C, if and only if* $\text{REDNF}(c, \langle i\rangle + I_C) = 0$.

Remark 5.2.10. *A Boolean Gröbner basis of $\langle i\rangle + I_C$ for a general i can be hard to compute. However in practice, often it can be constructed by adding i to the set E_C. One example is the case, when i has a linear leading term not present in E_C. A slightly more advanced case is the case that no variable in i occurs as leading term in E_C. Then we just have to compute a Boolean Gröbner basis of $\langle i\rangle$ and add it to the polynomials in E_C to obtain a Boolean Gröbner basis of $\langle i\rangle + I_C$.*

Counter examples for failed properties can be generated using algorithm 31. In contrast to algorithm 23, it computes just a single counter example. It needs as many time steps as variables in the polynomial ring and does not generate any polynomial intermediate expressions. So compared to the normal form computation before, the counter example can be generated in nearly zero time, if the result of the normal form is nonzero.

Algorithm 31 find_one: Find a point where a Boolean polynomial evaluates to one.

Input: Boolean polynomial $p \neq 0$
Output: variable configuration S, that after substitution of x by v for all $(x \to v) \in S$ in p one is obtained.

  if $p = 1$ **then**
    return $\emptyset$
  $x :=$ topvar(p)
  if $p^E = 0$ **then**
    $v := 1$ /* plug in 1 */
    $q := p^T$
  else
    $v := 0$ /* plug in 0 */
    $q := p^E$
  return $\{(x \to v)\} \cup$ find_one(q)

Theorem 5.2.11. *Algorithm 31 is correct and terminates.*

Proof. Termination is ensured by usual principle for the form of recursive functions on decision diagrams, see Remark 3.1.16. So let $p \neq 0$. If p is one, every point v will satisfy $p(v) = 1$. So fixing no coordinate of v and returning the empty set is just sufficient.

Now, we can treat the case, p is non-constant. Remember that using the notation of else- and then-branches (definition 3.1.15) p is of the form

$$p = \text{topvar}(p) \cdot p^T + p^E.$$

If p^E is 0, p is of the form topvar$(p) \cdot p^T$ ($0 \neq p^T$ a Boolean polynomial). Since p^T has a point where it evaluates to one and we can safely replace topvar(p) by 1. On the other hand, if p^E is not zero, we can just plug in 0 for topvar(x) and search for a one in p^E. □

Remark 5.2.12. *Splitting a property p into an if-condition i and a then part c, pulls out the factor $i + 1$. This should (we use Boolean multiplication) usually result in a smaller degree of the polynomial that is taken to compute the normal form.*

5.2.5 Symbolic Model Checking

Symbolic Model Checking is a method for verifying correctness of behaviour of automata. It is called symbolically, as the states are not represented explicitly (handling each state separately), but sets of them are considered: Usually this is done using BDDs; a single BDD can represent a huge set of states. The same holds for a system of polynomial equations: An ideal I represents a variety of states. So in principle, the term "Symbolic Model Checking" fits quite well to algebraic approaches.

The basic ideas for this connection have already been developed in Tran and Vardi [2007] and Avrunin [1996] on a rather theoretical basis.

While we do not present practical results here, the meaning of this section is to make "Symbolic Model Checking" as compatible as possible with topological ordering which is the precondition for getting results in reasonable time applying of techniques from computational algebra (Remark 5.2.3)

Before we describe how Gröbner bases can solve typical "Symbolic Model Checking" problems like reachability, we give a quite informal definition of automata (for a more sophisticated one we refer to Bérard et al. [1999]).

Definition 5.2.13. *A automaton is a sequential circuit. Signals derived from the previous time step are called* ***states****. Signals, which are stored for the next iteration, are called* ***next states****. Each next state variable will be transformed into a state variable in the next step (so we have a one to one correspondence here). Like other circuits it can have input signals which stimulate the circuit behaviour in each time step. The set of states is denoted by S, the set of next states N. Each state/next state signal s can have a different value s_t at time t.*

Remark 5.2.14. *In many reachability analyses existential quantification for the input signals is used (given our set of states/next states, it exists an input signal configuration, such that the state/next state can be reached/was reachable in the next/previous step). Set theoretically we project from the set of states, inputs and next states to the set of next states/(previous) states. Algebraically this means that we are interested in eliminating input and state/next state variables. In this description the first alternative always refers to the so called forward iteration and the second one to backward iteration. Forward iteration checks the reachability of future states, while backward iteration treats the question, what previous states could have lead into a specific set of states.*

Theorem 5.2.15. *Given an automaton, considering exactly one time step and the combinatorial circuit C cut out for this step, there exists a topological variable ordering on the signals of C, such that the state signals S form the last variable ordering in this variable ordering. Taking the lexicographical ordering with respect to this variable ordering we get an elimination ordering for all other variables.*

Proof. The exists no gate where the state variable forms an output. We can take any topological variable ordering and move the state variables at the end. □

In principle forward and backward iteration can be done by a series of Boolean Gröbner basis computations. The equations involved in the iteration at time t can be separated in two parts: I_C and the implicit representation of the set of state variables (forward iteration) S_t/next state variables (backward iteration) N_t.

Using a generating system or Boolean Gröbner basis of the ideal intersected with the Boolean polynomial ring in the set of next state/state variables, we get S_{t+1}/N_{t-1} for the next iteration step by mapping each next state/state variable to the corresponding state/next state variable.

Classical reachability analyses often consider an ascending chain (with respect to inclusion) of sets of states: Given an initial set of states, it is desired to know all states reachable in any number of finitely many iteration steps. That means that in each iteration step the newly computed set of states is joined with the previously computed sets. Since there exists only finitely many states of an automaton, the power set of these states is also finite, so in theory every ascending chain of state sets will stabilize at some time point t^*. The only theoretical bound for t^* is $2^{|S|}$ which can be quite high. In fact for some automata, this bound is sharp (you can imagine a clock storing the time in its state signals). On the other hand, in practice there exist automata where the iteration (with or without joining the previous state sets) stabilizes just after a few steps. This can depend very much on the example. Comparing two canonical generating systems (for example reduced Boolean Gröbner bases) equality of S_t and S_{t+1} can be determined quite efficiently.

Remark 5.2.16. *There exists an alternative to ideal intersections for getting these ascending chain of state sets. The state set can also be extracted from our polynomials using ZDD/BDD methods (for example determining the zeros of a polynomial in* PolyBoRi *is implemented quite directly on the ZDDs, algorithm 22). Then these sets can be manipulated using usual toolset for decision diagrams. Even where the use of BDDs for symbolic model checking fails, there is a good possibility that BDDs can help at this point. This is because in opposite to symbolic model checking using BDDs only $|S|$ instead of $|S| + |N| = |2 \cdot S|$ variables are involved in the intersection. This can reduce the complexity of diagram operations. Essentially for just comparing states sets it is not needed to reconvert them to polynomial ideals. In some cases it might increase the performance of Gröbner bases computations doing so. This can be considered in very opposite ways:*

- *Just take as state/next state ideal the previously computed next state/state ideal.*
- *Consider the union of all previous calculated state sets as input (state/next state).*
- *Consider last state set without every state that was previously in the input sets.*

The first method is the simplest and does not require any change of the ideal/state set. The other methods might improve performance, if the ideals involved in the calculation become simpler. From a theoretical point of view, it seems unclear which is the best way to calculate. So this is up to in-depth experiments.

Corollary 5.2.17. *Using theorem 5.2.15 we can use such an ordering to compute a Boolean generating system of $(I_C + N_t) \cap \mathbb{Z}_2[S]$ by the following step:*

1. $R := \{\mathrm{REDNF}(p, I_C) | p \in N_t\}$ *(E_C forms a Boolean Gröbner basis of I_C)*

2. *Compute a Boolean Gröbner basis B of R with respect to the ordering specified in theorem 5.2.15*

3. *The elements of $B \cap \mathbb{Z}_2[S]$ form the desired system.*

Proof. E_C is a Boolean Gröbner basis of I_C and its leading terms are variables not belonging to S. So they will not appear in the reduced normal forms any more. E_C together with B forms a Boolean Gröbner basis of $I_C + N_t$. Since our monomial ordering is an elimination ordering for the set of variables not belonging to S, we only need to take the elements of the Boolean Gröbner basis which lie in $\mathbb{Z}_2[S]$. Due to their leading terms, we know that elements of E_C do not belong to this category. □

Remark 5.2.18. *The forward iteration is harder to compute with Gröbner basis techniques, as in this case we cannot use a topological monomial ordering for the elimination. However it is possible to orient oneself as much as possible on topological orderings: For example you might consider a product ordering, in the first block restricting a topological ordering to all variables except N, in the second block a degree ordering on N. This technique already gave promising performance in first experiments. Another possibility is to* approximate *the result only. The elimination can be done using a product ordering and the Buchberger algorithm. Applying a degree bound to the Buchberger algorithm might yield less generators. So the ideal might become smaller and the variety bigger. In this way, we can get an upper bound for the set of states which might suffice for many applications.*

5.3 Conversion to CNF

Boolean polynomial form the so called **algebraic normal form (ANF)** of a formula in Boolean logic: Mapping 0 to False, 1 to True, + gets the meaning of XOR and $\star$ becomes logical and ($\wedge$). So an expanded (no brackets) Boolean polynomial represents an expression from propositional logic in terms of XOR and $\wedge$. In contrast to this algebraic normal form, the most used normal forms use logical and ($\wedge$) and logical or ($\vee$): the conjunctive and the disjunctive normal form. SAT-solvers usually expect a conjunctive normal form as input format. A variable or its negation is called a **literal**. A disjunction of literals is called a **clause**. The **conjunctive normal form** is a conjunction of disjunctions. Example:

$$(a \vee b) \wedge (\neg a \vee b \vee c) \wedge (b \vee \neg c)$$

The common algorithm used in cryptanalysis [Courtois and Bard, 2008] for converting from ANF to CNF introduces auxiliary variables. In contrast, we will be able to avoid additional variables, while still keeping the size of the CNF moderate.

In algorithm 32 we use the symbolic computation of the set of points where a polynomial evaluates to one. This was given in algorithm 23. We only have to introduce the algorithm for a single polynomial, as we can take the conjunction of the clauses generated for each single polynomial.

Definition 5.3.1. *Let $V \subseteq \mathbb{Z}_2^n$. A set B is called a* ***block*** *of V, if $B \subseteq V$ and there exist sets $\emptyset \neq C_1, \ldots, C_n \subseteq \{0,1\}$, such that*

$$B = \{(c_1, \ldots, c_n) | c_i \in C_i\}.$$

B is called a ***prime block*** *of V, if there exists no $B' \supsetneqq B$, such that B' is a block of V. For a formula f from propositional logic in n variables, we write* holds(f) *for the set of all points in $\mathbb{Z}_2^n$ where f evaluates to true.*

We construct the conjunctive normal form by finding prime blocks in the the set of points where our expression ($p = 0$) is not fulfilled. This is the usual approach for CNF generation as given in [Lipp, 2002]. For each block we store the negation. p evaluates to zero on a point v, if and only if p is not an element of any prime blocks. So the conjunction of the negation of the prime blocks forms a CNF of p.

The reason, why we can do this and why it was not done before (at least in algebraic cryptanalysis), is that we do not have to calculate an explicit truth-table. The size of the truth-table growths exponentially with the number of variables. This is not necessarily the case for the computation time and the size of the result of algorithm 23, as they only depend on the ZDD structure. In fact, for a typical polynomial, the algorithm will be quite fast.

Example 5.3.2.

$$\begin{aligned}
p &= x_1 \cdot x_2 + x_1 + x_2 \cdot x_3 + 1 \\
\text{ones}(p) &= \{(0,0,0), (0,0,1), (0,1,0), (1,1,0)\} \\
c_1 &= \{x_1, x_2\} \\
H_1 &= \{(0,0,0), (0,0,1)\} \\
c_2 &= \{\neg x_2, x_3\} \\
H_2 &= \{(0,1,0), (1,1,0)\} \\
\textit{result} &= (x_1 \vee x_2) \wedge (\neg x_2 \vee x_3)
\end{aligned}$$

Theorem 5.3.3. *Algorithm 32 is correct and terminates.*

Proof. **Termination:** The inner for-loop is invoked r-times. The outer while-loop terminates as T is finite and decreases in each step. So termination is clear.
Correctness: We denote by H_i, c_i the sets H, c in the i-th iteration of the while-loop after the end of the for-loop. Then we have

$$O = \{v \in \mathbb{Z}_2^n | p(v) = 1\} = \bigcup_i H_i.$$

Algorithm 32 ANF to CNF conversion

Input: p a Boolean polynomial

Output: $R \subseteq \mathcal{P}(\{x_1, \neg x_1, x_2, \ldots, \neg x_n\})$:$\bigwedge_{c \in R} \left(\bigvee_{l \in c} l\right)$ is a CNF of $p = 0$

```
  O := ones(p, Z_2^n)
  T := O
  R := ∅
  i := 0
  while T ≠ ∅ do
    i := i + 1
    choose o ∈ T
    H := {o}
    for j ∈ {1, ..., n} do
      c := ∅
      H' := {h|h ∈ H} ∪ {h + e_j|h ∈ H}
      /* try to enlarge the set by adding the j-th unit vector to each element */
      if H' ⊆ O then
        H := H'
      else
        if o_j = 1 then
          c := c ∪ {¬x_j}
          /* introduce the opposite of j-th entry of o in our clause c */
        else
          c := c ∪ {x_j}
      if c ≠ ∅ then
        R := R ∪ {c}
    H_i := H
    c_i := c
    T := T\H
  return R
```

Variables		Eqs.	Variables CNF		Clauses		CNF size (MB)		time (s) MiniSat2	
all	p		class.	alg. 32	class.	alg. 32	class.	alg. 32	class.	alg. 32
100	5	1000	6488	100	67150	8114	1.3	0.1	0.2	0.1
100	6	1000	8693	100	99547	15299	1.9	0.3	0.3	0.3
100	7	1000	10945	100	139511	29001	2.8	0.6	0.7	1.1
200	5	2000	14543	200	133968	16192	2.7	0.3	0.5	0.1
200	6	2000	20373	200	201289	30659	4.3	0.6	1.1	0.5
200	7	2000	26297	200	278875	57572	6.1	1.3	17.9	2.6
500	5	5000	39135	500	332344	40293	7.2	0.7	1.4	0.1
500	6	5000	56419	500	499742	75899	11.2	1.5	14.8	1.5
500	7	50000	447811	500	6928026	1439399	172.6	34.0	425.3	629.5
1000	5	10000	80803	1000	668490	80403	14.8	1.4	1.9	0.3
1000	6	10000	118008	1000	998960	151924	23.2	3.1	273.8	3.6
1000	7	100000	1089021	1000	13866038	2881932	358.3	69.7	38788.6	1383.8
2000	5	20000	164105	2000	1336792	161627	31.2	3.1	3.7	0.6

Table 5.4: CNF conversion for random polynomial systems. We compared the number of clauses and number of variables in the generated CNF representations using the classical approach by Gregory Bard with the new one presented in algorithm 32. The conversion results were processed in the solver MiniSat2.

Moreover, it holds for all i that

$$H_i = \{(h_1, \dots, h_n) | h_i \in \{0,1\}, \text{ if } x_i \in c_i : h_i = 0, \text{ if } \neg x_i \in c_i : h_i = 1\} = \mathbb{Z}_2^n \backslash \left(\bigvee_{l \in c_i} l \right).$$

This gives us, for the set of points using the final value of R, were $p = 0$ is fulfilled:

$$\{v \in \mathbb{Z}_2^n | p(v) = 0\} = \mathbb{Z}_2^n \backslash O = \bigcap_i (\mathbb{Z}_2^n \backslash H_i) = \bigcap_i \text{holds}(\bigvee_{l \in c_i} l) = \text{holds}(\bigwedge_{c \in R} \left(\bigvee_{l \in c} l \right))$$

□

We want to remark for the implementation that we can restrict ourselves to the treatments of variables really occurring in p. In Table 5.4 this number is listed in the second column. We used random polynomials in degree two which are dense (every term occurs with probability 0.5) in a limited subset of variables. The reason for this construction is that in computational algebra and algebraic cryptanalysis, a random, generic polynomial is very untypical. Moreover, we choose overdetermined systems. This does not influence the conversion process, but helps to construct systems which are solvable by the used SAT-solver (MiniSat2 [Eén and Sörensson, 2003]) in reasonable time. We compared an implementation [Albrecht, 2009] of the classical approach given in [Courtois and Bard, 2008] with our implementation of algorithm 32. The conversion time was moderate for both algorithms. Both conversion algorithms were implemented as Python scripts and are not yet optimized. All instances were satisfiable. The experimental data show that we were able to reduce the size of the CNF considerably in the presented examples.

Of course, there exists aspects which are orthogonal to this approach: It is still possible, as done in classical converters, to split long XOR-chains. Moreover, it seems to be unwise to pass any linear equation to the SAT-solver. These should be used to eliminate variables by normal forms. The values of these eliminated variables can be recomputed later, when the example turns out to be satisfiable. An alternative and probably preferable approach consists in the use of SAT-solvers which support the XOR-chains directly to encode linear polynomials [Soos et al., 2009].

The following two properties of algorithm 32 conclude this chapter.

Theorem 5.3.4. *The algorithm 32 converts a polynomial p of the form*

$$p = \prod_{k=1}^{r} (x_{i_k} + c_k),$$

where x_{i_k} are disjoint variables and c_k are constants, into the clause

$$\bigvee_{k=1}^{r} (x_{i_k} + c_k + 1).$$

Therefore, using algorithm 32 for converting a (disjunctive) clause into ANF (the natural way) and back is the identity on the set of clauses. Following, we will call such a Boolean polynomial a ***polynomial in clause form****.*

Proof. W.l.o.g. p involves all variables: $n = r$ and $(x_{i_1}, \ldots x_{i_r}) = (x_1, \ldots, x_n)$. Then p has a single one, namely $(c_1+1, \ldots, c_n+1)$. Hence the condition $H' \subseteq O$ will always fail. For $o_j = c_i + 1 = 1$ we add $\neg x_i = x_i + c_i + 1$ to the clause. This formula also holds for the case $o_j = 0$. Moreover, since T is initialized with the single one and shrinks in each iteration, only one clause is generated. □

Theorem 5.3.5. *Let I be an ideal with $\langle x_1^2+x_1, \ldots, x_n^2+x_n \rangle \subset I \subseteq \mathbb{Z}_2[x_1, \ldots, x_n]$. Let F be the reduced Boolean Gröbner basis of I for a given monomial ordering. If there exists $p \in I$ in clause form and a Boolean polynomial $f \in F$ with $\mathrm{lm}(f) = \mathrm{lm}(p)$, then $f = p$. In other words: Reduced Boolean Gröbner bases contain polynomials in clause form, if their leading terms are minimal in the ideal.*

Proof. Let $p = \prod_{k=1}^{r} (x_{i_k} + c_k)$, then each term t in the tail(p) divides $\mathrm{lm}(p)$. Since $\mathrm{lm}(p)$ is a leading monomial in the minimal Gröbner basis, no other monomial in the leading term divides it. Since t divides $\mathrm{lm}(p)$, we get that $t \notin \mathrm{L}(I)$ and tail(p) is reduced. So, p is contained in some reduced Gröbner basis. Since the reduced Gröbner is unique and $\mathrm{lm}(p) = \mathrm{lm}(f)$, we get $p = f$. □

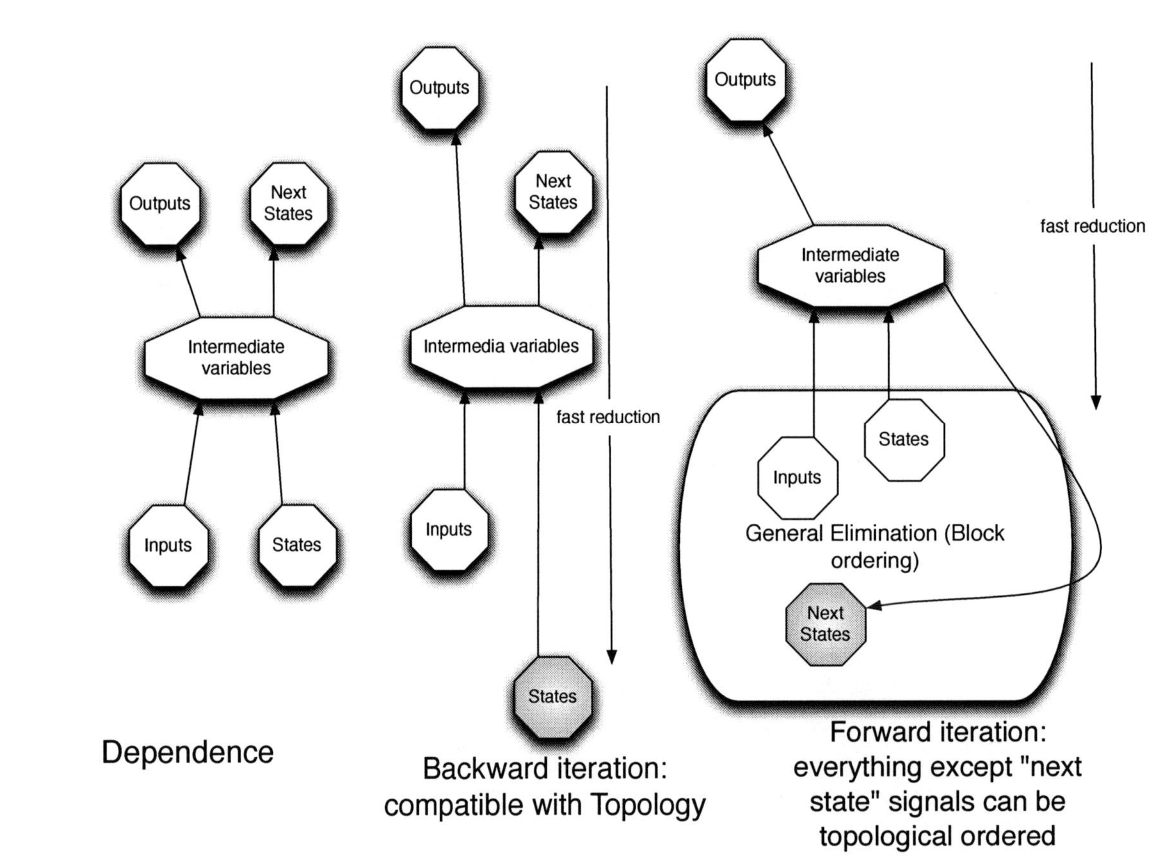

Figure 5.1: Forward and backward-iteration

Chapter 6

PolyBoRi as software

This chapter gives an overview of technical aspects of computing with ZDDs using the CUDD library and the PolyBoRi framework for computing Gröbner bases of ideals generated by Boolean polynomials and field equations.

6.1 Data structures

While PolyBoRi features many algorithmic advances, we started with the idea of computing Gröbner bases with ZDDs. To be precise, we were thinking about decision diagrams in general.

Of course, whenever someone tries something new it is obligatory to learn from the past. Our experience with Singular gave us much familiarity with different types of data structures. Actually, there are at least four completely different types of structures in Singular used for polynomials or polynomial systems:

1. ordered linked listed (the default structure, [Bachmann and Schönemann, 1998]),
2. geobuckets (see Yan [1998]),
3. recursive polynomial structures in factory, and
4. polynomials as rows in dense matrices (in FGLM, [Faugère et al., 1993, Wichmann, 1997]).

We observe that Singular uses several types of data structure for polynomial data during Gröbner basis computations: linked list, geobuckets, and matrices. One can conclude that each of them has its domain of application where it fits best and complex algorithms have to combine these different structures in an appropriate way. Another important point is that structures and algorithms depend on each other. If the complexities of some low-level operations change (some increase, some decrease), this has effect on algorithms. In PolyBoRi we use polynomials implemented on top of ZDDs. For monomials we have two representations: a ZDD-based and one using vectors from the standard template library [Stepanov and Lee, 1994]. We will call the latter BooleExponent. The class BooleExponent was introduced for the following reason: When dealing with a lot of single monomials the overhead of reference counting in ZDD computations,

	ZDD	**linked list**	**geobuckets**[1]	**factory**[2]	**dense**
leading term	++/0/ − −[3]	++	+	+	0
p1 + p2	+	+	++	0	+
tail(p)[4]	−	++	+	0	++
redNF(p,m)	++	0	−	− −	0
canonical	++	+	− −	+	+
mem. use (sparse)	++[5]	+	−	− −	− −
mem. use (dense)	0[5]	− −	− −	− −	++
mem. use (structured)	++[5]	−	−	− −	−
mem. use (many vars.)	++	−	−	−	− −
mem. use (few vars.)	− −	++	++	−	++
iteration	0[6]	++	− −	−	0
p1 * p2	++[7]	+	− −	+	0
factorize	++	−	− −	++	− −

[1]) partly reduced container   [2]) recursive expression
[3]) lex — special cases — general case   [4]) often during reduction with monomial
[5]) every substructure stored once
[6]) possible, but ordering-dependent   [7]) automatically reduced

Table 6.1: Comparison of Data Structures

can become quite big. Also vectors are much more compact than ZDDs in the case of monomials since ZDDs look like linked lists with reference counters when representing single terms (see Figure 3.6(c)). As counter part to our ZDDs we use dense matrices and linear algebra. This structure has fundamentally different properties than a sparse polynomial representation. Hence, it can supplement our ZDD structures in many examples. In general, the memory use on the respective polynomials can be a good hint which data structures are suitable.

6.1.1 Algebraic structure

One property, that makes the polynomial term structure in ZDDs (or equivalently the Boolean function in a FDD) particularly suited for computational algebra, is that it is transparent for operations in computational algebra. Monomials correspond to paths in ZDD. The natural path order corresponds to the lexicographical monomial ordering (section 3.3.1). A very efficient memory storage is the compression of a dense polynomial structure using algorithms like gzip [Ziviani et al., 2000]. This has already been tried in [Faugère, 1999]. Of course, such a method hides the term structure completely. So every algebraic operation needs a decompression.

Also, we store polynomials as decision diagrams. We tried to avoid strictly comparing every single term of a polynomial for getting the leading term. This would yield an unacceptable performance bottle neck when computing Gröbner bases. Although we are not able to achieve this goal in general monomial orderings, it is quite possible for simpler, well selected orderings (section 3.3). For the degree reverse lexicographical ordering we had to change the order of variables (compared to the convention in the SINGULAR computer algebra system) which can be compensated by reversing the variables in the input system.

6.1.2 Canonicity

We can distinguish different levels of canonicity of data structures. One of the most uncanonical forms of polynomials is their representation as Geobuckets. In this form polynomials are represented as the sum of several (possibly) smaller polynomials. However, this is not a unique representation. For example, the equation

$$(x+y)+(z)=(x)+(y+z)=(x+y+z)+0=(2x+y+z)+(-x)$$

gives us four different representations of the same polynomial. Actually, there exist infinitely many in characteristic 0. More canonical is the representation as ordered linked list of terms. In this case, equality can be tested term by term.

Much more interesting than simple equality testing is the availability of hash functions [Knuth, 1998]: A hash function maps a data structure (like polynomials) to a hash value. It is stored in a small and easy comparable data type like `int`. A good hash function tries to separate the values as good as possible: mapping distinct polynomials to distinct hashes. While it is in general (e. g. hashing character strings) not always possible to achieve unique hash values (injective hash function), this is indeed possible in the case of ZDDs using the CUDD library. On most architectures there exist much more ZDDs than hash values. However these can not be represented in the virtual memory at the same time. Up to now the argument would also hold for linked list polynomials in SINGULAR. However, in the latter case, there exists no fast way to compute this hash value. In the case of ZDDs their internal data structure has a unique memory address: No ZDDs are stored twice. So, we can essentially use this pointer as hash value which is the fastest possible way to compute a hash value. We do not even have to investigate the graph structure of the diagram. Having these hash values, it is possible to implement very fast associative containers: structures mapping from polynomials to other data types (instead of integers they can be indexed by polynomials). These maps are key components in our Gröbner bases algorithms, as they facilitate the access to Boolean generators by their leading term (see section 4.4.1).

6.1.3 Functional programming

The CUDD library, which provides the low level functions for the ZDDs in POLYBORI, follows a functional programming style (Abelson et al. [1996]). While the underlying programming language C does not follow this paradigm, it provides a mechanism for handling ZDDs which enforces a style found in many functional languages like Lisp or Haskell [Hudak, 2000]. Functional programming is reference transparent: There are no operations changing data, it is only possible to create new objects. In this way functional programming feels often quite natural for mathematical concepts, but quite strange in contexts where we do not think in a functional way: for example in input and output on the terminal.

It is very important to notice that the functional style leverages other properties in the CUDD library like reference counting, the uniqueness of ZDDs (6.1.2), caching, and recursive programming (usually the smallest index of the arguments increases in each recursion step).

In principle functional programming is well suited for multi-threading (distributing a task on several CPU cores) [Hammond, 1994]. Unfortunately the CUDD library does not provide any support for this (except using different managers/polynomial rings). This will be a challenge in future to incorporate full support for multiple CPUs in PolyBoRi. The M4RI library, which is the backend for linear algebra in PolyBoRi, supports multiple cores using OpenMP [Dagum and Menon, 1998]. However it would be nice to have this ability also on the level of polynomials/ZDDs.

6.1.4 Manipulatability

When manipulating data structures we distinguish between destructive and non-destructive operations (changing or preserving their arguments). Destructive operations have the advantage of being fast, but they have side effects which are incompatible with functional programming. Linked list as they are used in Singular are highly manipulatable. This can result in very fast operations, but at the same time it causes often headaches debugging unwanted side effects of these very smart tricks. Despite the fact, that operations in PolyBoRi always create new ZDDs (maybe sharing some nodes of the original ones), the ability to create new diagrams involving nodes from existing diagrams in CUDD features a good expressiveness; when combined with caching it provides good performance. This can be seen as an analogy to Haskell programs: A literal translation in other programming languages results in horrible performance. But (while quite mathematically formulated) Haskell programs are very fast since the caching is integrated in the language. This has the consequence that the functional style is strictly enforced. Of course, there are operations where the overhead of functional programming is really high.

6.2 The PolyBoRi framework

Often we are asked what PolyBoRi exactly is. PolyBoRi is a software composed of several pieces. It is written in both: Python and C++. The code written in Python tends to be of quite high level and calls the C++ core routines. The glueing between C++ and Python is done using Boost::python which provides a very natural and sophisticated API for binding C++ libraries to Python.

PolyBoRi is natively built on high quality software components:

1. CUDD (ZDDs) [Somenzi, 2005],
2. several part of the Boost framework [Abrahams and the Boost Team, 2009],
3. libM4RI (asymptotically fast linear algebra in $\mathbb{Z}_2$) [Albrecht and Bard, 2008],
4. Python [Rossum and Drake, 2006, Martelli et al., 2005],
5. graphviz (optional for plotting) [Gansner, 2003], and
6. pyprocessing (optional for parallelization) [Oudkerk, 2009],

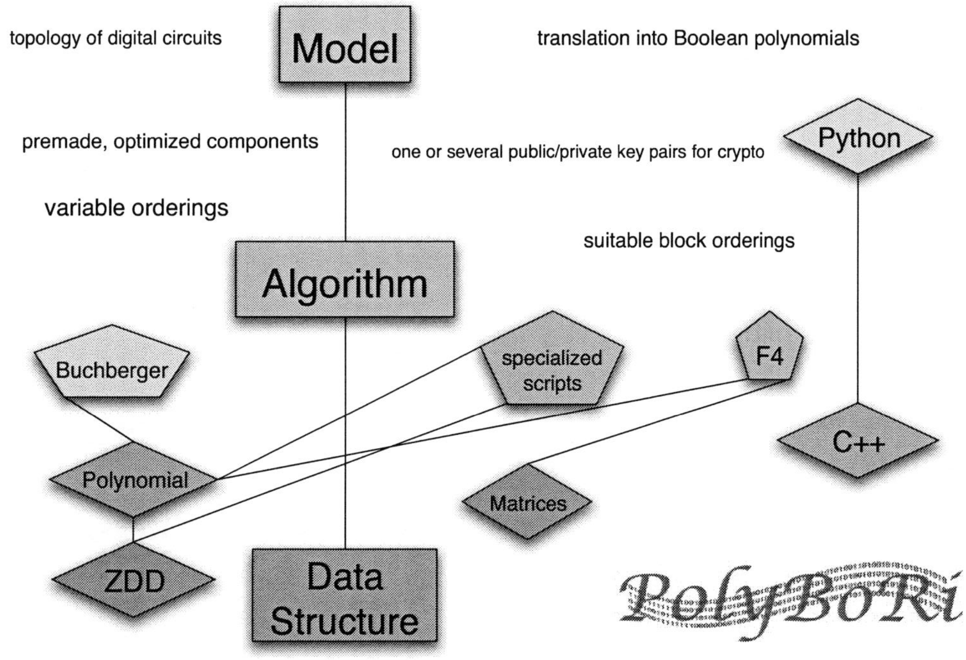

Figure 6.1: PolyBoRi- a Framework

7. SQLAlchemy (optional, object relational mapping) [Bayer, 2009]

8. sqlite (optional, database) [Hipp et al., 2009].

We have a public mercurial [Mackall, 2009] repository for the PolyBoRi source code at [Brickenstein and Dreyer, 2009b]. We include the source code statistics for PolyBoRi obtained from `http://www.ohloh.net` [Allen and Collison, 2004] in Figure 6.2 and Figure 6.3. The numbers present the state of the September 24th 2009.

Language	Code Lines	Comment Lines	Comment Ratio	Blank Lines	Total Lines
C++	221,077	12,328	5.3%	8,705	242,110
C	69,679	28,300	28.9%	22,021	120,000
HTML	30,257	135	0.4%	6,267	36,659
Python	5,880	1,322	18.4%	1,157	8,359
TeX/LaTeX	1,067	62	5.5%	265	1,394
Make	637	364	36.4%	278	1,279
shell script	241	113	31.9%	104	458
CSS	166	10	5.7%	18	194

Figure 6.2: Programming language statistics for PolyBoRi [Allen and Collison, 2004], September 24th, 2009

Regarding the statistics of the source code, we would like to remark that the C-Code consists mainly out of the distributed libraries M4RI and CUDD. The C++ code is mainly part of the PolyBoRi project itself. However, the enormous amount of 220000 lines of C++-code also contains 197000 lines of tables with Gröbner bases. Not included in this numbers is the PolyBoRi test suite. We run about 1000 tests every night and 10000 tests per weekend. Statistics and timings of these tests are stored in a sqlite database. While we did these regular tests for correctness and performance regression from the very beginning, the database containing the timings for better analysis is quite new. Between May and October 2009 it grew to over 100MB. This makes it possible to track the performance development of each example for different settings over time.

The PolyBoRi framework implements a fast and flexible Gröbner basis algorithm as well as the tools building new specialized algorithms. These include not only polynomial

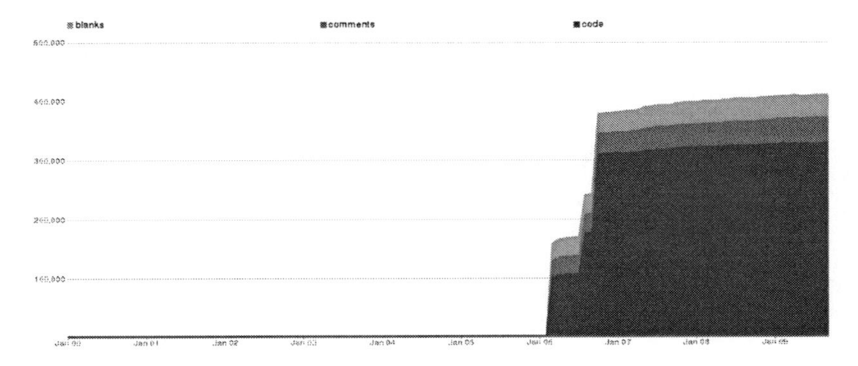

Figure 6.3: Lines of code in the PolyBoRi repository [Allen and Collison, 2004], September 24th 2009

arithmetic, but also building blocks of Gröbner basis algorithm (like normal forms, handling of critical pairs, criteria). We are strongly convinced that most problems in Boolean rings are quite special and that no prebuilt algorithm will ever provide the capabilities to compute them all. So, we would like to give researchers the opportunity to build an algorithm for their problem using our optimized tool chain.

In our own experiments we often have been able to handle problems involving a large number of variables and equations using such special scripts. On the other hand, we tried to improve the general implementation for computing Gröbner bases. In many cases these tricks can be seen as pre- and post processing filter for the Buchberger algorithm. This is implemented using Python decorators that add single aspects [Kiczales et al., 1997] to the core Buchberger algorithm. We tried to integrate as many of these ideas as possible in our main `groebner_basis` function to provide a large variety of algorithmic variants.

But, we not only optimize the algorithms. The best chances for improving performance lie on the side of modeling and algebraic formulation. Problems can be formulated in quite different ways: Even the Boolean variables can represent quite different objects. For example, in the popular Sudoku problems there exist several ways to encode the relation that an integer variable x takes values in $\{1, \ldots, 9\}$ in Boolean variables. For instance,

1. for each value i in $\{1, \ldots, 9\}$ define x_i to be 1 if and only if $x = i$, or

2. encode the value of x in four Boolean variables representing the bits of x.

Additionally, there have to be added further equations or compatibility conditions to model the Sudoku rules (e.g. in the first case prevent that $x_i = 1$ and $x_j = 1$ if $i \neq j$). These gives a completely different equation system (the second one has only $\frac{4}{9}$ as many variables as the first one and seems to give much better performance in our practical experiments).

Moreover there have been other formulations of the Sudoku problems with Gröbner bases:

- Sato, Suzuki, Inoue, and Nabeshima [2008b] formulated the equations as polynomials over $\mathbb{Z}_2^9$. Using modules these equations can be translated to $\mathbb{Z}_2$. This will result in a complete new system of equations, again with new meanings of the variables.
- Gago-Vargas, Hartillo-Hermoso, Martín-Morales, and Ucha-Enríquez [2006] give equations over the rational numbers.

Also you can optimize the formulation of equations in at least four different ways:

1. overdetermine the system as much as possible,
2. minimize their degree,
3. minimize their number of terms, and
4. precompute Gröbner bases for subsystems (often there are many subsystems symmetric[1] to each other).

We also tried to provide good facilities for the modeling integrated circuits verification problems (many of them are also useful for other domains):

1. A very high level inner domain specific language for specifying algebraic equation systems using predefined building blocks. An example is given in Figure 6.4.
2. A basic library for building blocks of integrated circuits (adder blocks, shift units).
3. Support for building block hierarchies.
4. Automatically generated hints for a block structure derived from these building blocks.
5. A converter from GAT-files to the PolyBoRi data format (however the GAT-format does not support XOR natively which leads to bad encoded circuits from the algebraic point of view).

[1]in the sense of section 2.2.2

```
adder_bits=256
adder_block=AdderBlock(sums="s",carries="c",
    input1="a",input2="b",
    adder_bits=adder_bits,start_index=1)
#input blocks ascending by bit index
#sum, carry, input a, input b

declare_ring([Block("x",100),adder_block,Block("y",100)])
ideal=[
    y(0)+y(1)+1,
    a(adder_bits)+1,b(adder_bits)+y(0)+y(1),
    x(0)+(c(adder_bits))*x(1)
]

adder_block.implement(ideal)

claims=[
  c(adder_bits)+1,
  if_then([x(1)],[x(0)]),
  if_then([a(1),b(1)],[c(1),s(1)]),
  if_then([a(1)+1],[c(1)+b(1),s(1)+b(1)+1]),
  # if-part a list of linear polynomials or
  # polynomials with linear lead
  s(adder_bits)+c(adder_bits-1),
  if_then([],[x(33)],supposed_to_be_valid=False)
]
```

Figure 6.4: PolyBoRi data format language

6.3 Comparison with classical computer algebra systems

Following, we briefly describe recent advances achieved using POLYBORI. We have concentrated ourselves on problems that could not be solved by algebraic methods before. For this purpose, we have made extensive use of the facilities of the POLYBORI framework.

1. In section 5.1.5 we showed that our interpolation algorithm can compute lexicographical normal forms without using a Gröbner basis for large random varieties (e. g. 500000 points). Moreover, we were able to calculate the Gröbner bases of the respective ideals for these random varieties. Even more, because representing the variety in a ZDD, we were able to handle even larger sets in the case of structured varieties. Our largest example involved 2501 variables and had more than 10^{308} interpolation points. Clearly, those examples are not representable in classical computer algebra systems.

2. State of the art for checking equivalence of multipliers with algebraic methods was about 4×4 bits, before we introduced topological orderings.

   In [Brickenstein, Dreyer, Greuel, Wedler, and Wienand, 2009] we have experienced that even using these orderings 6×6 is the bound for classical computer algebra systems while we were still able to go up to 10×10 with the general routines of POLYBORI. In section 4.3, we have introduced algorithms for computing the needed normal form in one recursive computation with ZDDs. Using these techniques in POLYBORI we lifted the bound to 16×16. Together with the optimizations from section 5.2.2 and section 5.2.3, we were able to get a time of 24 seconds on a MacBook Pro 2.5 GHz Core Duo 2 (using a single core only). In Table 6.2 we also compared the number of possible input vectors for which we verified the integrated circuit with the number of seconds needed for the calculation. For a 16×16 multiplier the symbolic computations verify the behaviour for all theoretic $2^{32} = 4294967296$ values of input vectors.

   An almost constant quotient between the number of input vectors and computing time would be typical for a simulator and result in an exponential complexity of the algorithm. While we do not have any theoretical results about the asymptotic complexity of our method, it is a promising sign that this quotient growths with a factor of 200 between the 8×8 and the 16×16 examples. We like to repeat at this point that the theoretical results in [Becker et al., 1995] about exponential size of FDDs for multipliers are not immediately applicable here, because in the case of equivalence the resulting normal form is just 0.

Example	time	#inputs	#variables	(input vectors)/seconds
multiplier 8x8	0.07	16	204	940000
multiplier 10x10	0.14	20	314	7500000
multiplier 16x16	23.65	32	768	180000000

Table 6.2: Equivalence check of multipliers using POLYBORI

As Table 6.2 indicates, it is highly probable that we are able to handle even larger circuits. Unfortunately, larger benchmark examples are not available in a suitable input format up to now. Finally, we would like to remark that this work is in no way specific for multipliers, but applicable to the bit-level verification of general integrated circuits.

3. In [Bulygin and Brickenstein, 2010] we improved the state of the art in algebraic cryptanalysis. We were able to attack small scale AES cipher from 8 bits to 64 bits for two rounds [Cid et al., 2005]. This also involved theoretical improvements of the attack. From a computational point of view, the new introduced method for linear lexicographical leading terms contributed substantially to the result (section 4.3).

4. In Table 4.1.3 we showed basic arithmetic with huge, random polynomials which are too big for classical computer algebra systems. Our largest example involved two polynomials in 21 variables, each with 250000 terms. We were able to multiply them in five seconds.

5. Also external researchers were able to improve their results substantially using the PolyBoRi software. On the one hand, Gligoroski, Dimitrova, and Markovski [2009] used it for interpreting quasi-groups as vector valued Boolean functions. On the other hand, Albrecht and Cid [2009] used our software to combine algebraic and statistical techniques in cryptanalysis. Finally, Guerrini, Orsini, and Simonetti [2009] claimed that computing distance distributions with PolyBoRi had a better *complexity* than using the Magma [Bosma et al., 1997] or Singular [Greuel et al., 2009].

6. Gao [2009] presented some example sets of CNF instances, that could not be solved by several compared DPLL-implementations in reasonable time or in time comparable to PolyBoRi's Gröbner bases facilities. Using PolyBoRi, he was able to solve the #SAT problem [Gomes et al., 2009], i. e. counting the solutions of an equation system. Mathematically he accomplished this task using the equality between the number of standard monomials of radical ideals and solutions over the ground field (theorem 5.1.6). Regarding the counting of standard monomials, we can improve his results by reduction of the problem to counting of valid paths in ZDDs: By using the structure of the diagram very large structured sets can be counted without enumerating every single element.

Moreover, we would like to point out that an independent comparison of PolyBoRi with other systems (including their own) was done in [Gerdt and Zinin, 2008]. PolyBoRi won most of the examples, some even at a very high factor of 100 or more[2].

[2]Regarding the methodology of the benchmarks, we would like to add that about $0.5s$ should be subtracted from every PolyBoRi timing, because they included our startup time. Moreover, it should be noted that the full potential of PolyBoRi is much higher, as the example data included only a very small number of variables (typically 10-20) while the PolyBoRi data structures were constructed for much larger variable sets.

Chapter 7

PolyBoRi Tutorial

This section gives a tour through the PolyBoRi framework. The major focus is set on the ZDD data structures: We will try to give some insights in them which hopefully will help to write performant programs using PolyBoRi.

7.1 Introduction

This section gives some overview of the PolyBoRi framework and shows first usage examples. The following section will go more in detail about the functionality presented in this section.

7.1.1 Interfaces

The core of PolyBoRi is a C++ library. On top of it there exists a Python interface. Additionally a Python-based integration into SAGE [Stein et al., 2009] was provided by Burcin Erocal and Martin Albrecht. The main difference is that PolyBoRi's built-in Python interface makes use of the boost [Abrahams and the Boost Team, 2009] library while the SAGE interface relies on Cython [Behnel et al., 2009]. However the wrappers for SAGE and the original Python interface are designed that it is possible to run the same code under both bindings.

We provide an interactive shell for PolyBoRi using IPython [Perez and Granger, 2007] for the SAGE interface (which is invoked the command `sage`) as well as for the built-in one which can be accessed by typing `ipbori` at the command line prompt.

In ipbori a global ring is predefined and a set of variables called $x(0)\ldots,x(9999)$. The default ordering is lexicographical ordering (lp).

7.1.2 Ring declarations

While in ipbori usually a standard ring is predefined, it is possible to have more advanced ring declarations using the `declare_ring`-function which takes two arguments. The first argument is a list of variables or blocks or variables, and the second is a dictionary where the ring and the variable declarations are written to. The ring is always named `r` in the given dictionary (a typical choice is to use the dictionary of global variables). In addition to that the ring is returned.

Example

```
In [1]: declare_ring([Block("x",2),Block("y",3)],globals())
Out[1]: <polybori.dynamic.PyPolyBoRi.Ring object at 0x18436f0>

In [2]: r
Out[2]: <polybori.dynamic.PyPolyBoRi.Ring object at 0x18436f0>


In [3]: [Variable(i) for i in xrange(r.n_variables())]
Out[3]: [x(0), x(1), y(0), y(1), y(2)]

In [4]: declare_ring([AlternatingBlock(["x","y"],2)],globals())
Out[4]: <polybori.dynamic.PyPolyBoRi.Ring object at 0x2370b70>



In [5]: [Variable(i) for i in xrange(r.n_variables())]
Out[5]: [x(0), y(0), x(1), y(1)]


In [6]: [Variable(i) for i in xrange(r.n_variables())]
Out[6]: [x(0, 0), x(0, 1), x(0, 2), x(1, 0), x(1, 1), x(1, 2)]

In [7]: declare_ring(["x","y","z"],globals())
Out[7]: <polybori.dynamic.PyPolyBoRi.Ring object at 0x2eb4630>

In [8]: [Variable(i) for i in xrange(r.n_variables())]
Out[8]: [x, y, z]
```

7.1.3 Ordering

The monomial ordering can be changed by calling `change_ordering(code)` where `code` can be either `lp` (lexicographical ordering), `dlex` (degree lexicographical ordering), `dp_asc` (degree reverse lexicographical ordering with ascending variable ordering), `block_dlex` or `block_dp_asc` (for ordering composed out of blocks in the corresponding ordering). When using block ordering, after changing to that ordering, blocks have to be defined using the `append_ring_block` function.

In contrast to the lexicographical, degree lexicographical ordering, and the degree reverse lexicographical ordering in SINGULAR, our degree reverse lexicographical ordering has a reverse variable order (the first ring variable is less than the second, the second less than the third). This is a result of the fact that efficient implementation of monomial orderings using ZDD structures is quite difficult (and the performance depends on the ordering).

Example

```
In [1]: r=declare_ring([Block("x",10),Block("y",10)],globals())
```

```
In [2]: x(0)>x(1)
Out[2]: True

In [3]: x(0)>x(1)*x(2)
Out[3]: True

In [4]: change_ordering(dlex)

In [5]: x(0)>x(1)
Out[5]: True

In [6]: x(0)>x(1)*x(2)
Out[6]: False

In [7]: change_ordering(dp_asc)

In [8]: x(0)>x(1)
Out[8]: False

In [9]: x(0)>x(1)*x(2)
Out[9]: False

In [10]: change_ordering(block_dlex)

In [11]: append_ring_block(10)

In [12]: x(0)>x(1)
Out[12]: True

In [13]: x(0)>x(1)*x(2)
Out[13]: False

In [14]: x(0)>y(0)*y(1)*y(2)
Out[14]: True

In [15]: change_ordering(block_dp_asc)

In [16]: x(0)>x(1)
Out[16]: False

In [17]: x(0)>y(0)
Out[17]: False

In [18]: x(0)>x(1)*x(2)
Out[18]: False
```

```
In [19]: append_ring_block(10)

In [20]: x(0)>y(0)
Out[20]: True

In [21]: x(0)>y(0)*y(1)
Out[21]: True
```

In this example we have an ordering composed of two blocks, the first with ten variables, the second contains the rest of variables $y(0), \ldots y(9)$ (per default indices start at 0).

Even, if there is a natural block structure like in this example, we have to explicitly call `append_ring_block` in a block ordering to set the start indices of these blocks.

This can be simplified using the variable `block_start_hints` created by our ring declaration.

```
In [1]: declare_ring([
   ...:    Block("x",2),
   ...:    Block("y",3),
   ...:    Block("z",2)],
   ...:    globals())
Out[1]: <polybori.dynamic.PyPolyBoRi.Ring object at 0x1848b10>

In [2]: change_ordering(block_dp_asc)

In [3]: for b in block_start_hints:
   ...:     append_ring_block(b)

In [4]: block_start_hints
Out[4]: [2, 5]
```

Another important feature in PolyBoRi is the ability to iterate over a polynomial in the current monomial ordering.

```
In [1]: r=declare_ring([Block("x",10),Block("y",10)],globals())

In [2]: f=x(0)+x(1)*x(2)+x(2)

In [3]: for t in f.terms():
   print t

x(0)
x(1)*x(2)
x(2)

In [4]: change_ordering(dp_asc)

In [5]: for t in f.terms():
    print t
```

```
x(1)*x(2)
x(2)
x(0)
```

This is a nontrivial functionality as the natural order of paths in ZDDs is lexicographical.

7.1.4 Arithmetic

Recall that Boolean Polynomial polynomials are polynomials over $\mathbb{Z}_2$ where the maximal degree per variable is one. So Boolean polynomials are the canonical (reduced) representatives of the residue classes in $\mathbb{Z}_2[x_1, \ldots, x_n]/\langle x_1^2 + x_1, \ldots, x_n^2 + x_n\rangle$. Arithmetic in PolyBoRi can be considered as arithmetic between the residue classes. From a theoretical perspective, we always tried to avoid the quotient ring by considering ideals in the polynomial ring containing the field equations. For this reason in definition 2.1.3, we have introduced the notion of Boolean multiplication $\star$ between Boolean polynomials. The Boolean multiplication is equivalent to the one of the quotient ring, but takes place in the polynomial ring itself. If exponents bigger than one per variable appear reduction by the field ideal (polynomials of the form $x^2 + x$) is done automatically. While we have clear theoretical preference, regarding the PolyBoRi software it is left to the user which interpretation of the objects suits more to him.

```
In [1]: Polynomial(1)+Polynomial(1)
Out[1]: 0

In [2]: x(1)*x(1)
Out[2]: x(1)

In [3]: (x(1)+x(2))*(x(1)+x(3))
Out[3]: x(1)*x(2) + x(1)*x(3) + x(1) + x(2)*x(3)
```

7.1.5 Set operations

In addition to polynomials, PolyBoRi implements a data type for sets of monomials, called `BooleSet`. This data type is also implemented on the top of ZDDs and allows to see polynomials from a different angle. Also, it makes high-level set operations possible which are in most cases faster than operations handling individual terms because the complexity of the algorithms depends only on the structure of the diagrams.

Polynomials can be converted to `BooleSets` by using the member function `set()` in constant time.

```
In [1]: f=x(2)*x(3)+x(1)*x(3)+x(2)+x(4)

In [2]: f
Out[2]: x(1)*x(3) + x(2)*x(3) + x(2) + x(4)

In [3]: f.set()
Out[3]: {{x(1),x(3)}, {x(2),x(3)}, {x(2)}, {x(4)}}
```

One of the most common operations is to split the set into cofactors of a variable. This illustrates the following example.

```
In [4]: s0=f.set().subset0(x(2).index())

In [5]: s0
Out[5]: {{x(1),x(3)}, {x(4)}}

In [6]: s1=f.set().subset1(x(2).index())

In [7]: s1
Out[7]: {{x(3)}, {}}

In [8]: f==Polynomial(s1)*x(2)+Polynomial(s0)
Out[8]: True
```

The use of navigators is even more low-level than operation with `subset`-method. Navigators provide an interface to diagram nodes, accessing their index as well as the corresponding then- and else-branches (definition 3.1.15). Given a ZDD b, the top(b) can be accessed via `b.navigation().value()`.

```
In [1]:f=x(1)*x(2)+x(2)*x(3)*x(4)+x(2)*x(4)+x(3)+x(4)+1

In [2]:s=f.set()

In [3]:s
Out[3]:{{x(1),x(2)}, {x(2),x(3),x(4)},
     {x(2),x(4)}, {x(3)}, {x(4)}, {}}

In [4]:x(1).index()
Out[4]:1

In [5]:s.subset1(1)
Out[5]:{{x(2)}}

In [6]:s.subset0(1)
Out[6]:{{x(2),x(3),x(4)}, {x(2),x(4)}, {x(3)}, {x(4)}, {}}

In [7]:nav=s.navigation()

In [8]:BooleSet(nav,s.ring())
Out[8]:{{x(1),x(2)}, {x(2),x(3),x(4)},
     {x(2),x(4)}, {x(3)}, {x(4)}, {}}

In [9]:nav.value()
Out[9]:1

In [10]:nav_else=nav.else_branch()
```

```
In [11]:nav_else
Out[11]:<polybori.dynamic.PyPolyBoRi.CCuddNavigator
    object at 0xb6e1e7d4>

In [12]:BooleSet(nav_else,s.ring())
Out[12]:{{x(2),x(3),x(4)}, {x(2),x(4)}, {x(3)}, {x(4)}, {}}

In [13]:nav_else.value()
Out[13]:2
```

You should be very careful and always keep a reference to the original object when dealing with navigators, as navigators – very much like C++-iterators – contain only a raw pointer as data. For the same reason, it is necessary to supply the ring as argument when constructing a set out of a navigator.

The opposite of navigation down a ZDD using navigators is to construct new ZDDs in the same way, namely giving their else- and then-branch as well as the index value of the new node.

```
In [1]:f0=x(2)*x(3)+x(3)

In [2]:f1=x(4)

In [3]:if_then_else(x(1),f0,f1)
Out[3]:{{x(1),x(2),x(3)}, {x(1),x(3)}, {x(4)}}

In [4]:if_then_else(x(1).index(),f0,f1)
Out[4]:{{x(1),x(2),x(3)}, {x(1),x(3)}, {x(4)}}

In [5]:if_then_else(x(5),f0,f1)
---------------------------------------------------
<type 'exceptions.ValueError'>  Traceback (...)

/home/user/PolyBoRi/<ipython console> in <module>()

<type 'exceptions.ValueError'>: Node index must be less
   than top indices of then- and else-branch.
```

It is strictly necessary that the index of the created node is less than the index of the branches. Additionally to the use of navigators, operations on higher levels are possible using the set structure like the calculation of the minimal terms (with respect to division) in a `BooleSet`:

```
In [1]: f=x(2)*x(3)+x(1)*x(3)+x(2)+x(4)

In [2]: f.set()
Out[2]: {{x(1),x(3)}, {x(2),x(3)}, {x(2)}, {x(4)}}
```

```
In [3]: f.set().minimal_elements()
Out[3]: {{x(1),x(3)}, {x(2)}, {x(4)}}
```

7.1.6 Gröbner bases

Gröbner bases functionality is available using the function `groebner_basis` from polybori.gbcore. It has quite a lot of options and an exchangeable heuristic. In principle, there exist standard settings, but – in the case that an option is not provided explicitly by the user – the active heuristic function may decide dynamically by taking the ideal, the ordering and the other options into account which is the best configuration.

```
In [1]: groebner_basis([x(1)+x(2),(x(2)+x(1)+1)*x(3)])
Out[1]: [x(1) + x(2), x(3)]
```

There exists a set of default options for `groebner_basis`. They can be seen, but not manipulated, by accessing `groebner_basis.options`. A second layer of heuristics is incorporated into the `groebner_basis`-function, to choose dynamically the best options depending on the ordering and the given ideal. Every explicitly given option takes effect, but for the other options the default may be overwritten by the script. This behaviour can be turned off by calling

`groebner_basis(I,heuristic=False)`.

Important options are the following:

- `other_ordering_first`, possible values are *False* or an ordering code. In practice, many Boolean examples have very few solutions and a very easy Gröbner basis. So, a complex walk algorithm (which cannot be implemented using the PolyBoRi data structures) seems unnecessary as such Gröbner bases can be converted quite fast by the normal Buchberger algorithm from one ordering into another ordering. However, we have also implemented the FGLM algorithm for conversion of Gröbner bases.
- `faugere`, turn off or on the linear algebra [Faugère, 1999, Albrecht and Bard, 2008]
- `linear_algebra_in_last_block`, this affects the last block of block orderings and degree orderings. If it is set to *True*, linear algebra takes effect in this block.
- `selection_size`, maximum number of polynomials for parallel reductions
- `prot`, turn off or on the protocol
- `red_tail`, tail reductions in intermediate polynomials, this options affects mainly heuristics. The reducedness of the output polynomials can only be guaranteed by the option `redsb`
- `minsb`, return a minimal Gröbner basis
- `redsb`, return a reduced Gröbner basis (minimal and all elements are tail reduced)
- `clean_and_restart_algorithm`, from time to time restart the algorithm to clean the system of non minimal elements

7.1.6.1 Parallelization

With a view towards parallel computing PolyBoRi also has a preliminary support for concurrent processing of different strategies. For this purpose the function
`groebner_basis_first_finished` had been established. Very much like the command `groebner_basis` the first argument is assumed to be the system of equations which is given as a list of Boolean polynomials. Further arguments are the options for the different strategies (in form of dictionaries with keyword arguments for `groebner_basis`).

It returns the result of the variant which terminates first.

```
In [1]: from polybori.parallel import\
    groebner_basis_first_finished

In [2]: groebner_basis_first_finished([x(2)*x(1), x(1)+1],
   {'heuristic':False}, {'heuristic': True})
Out[2]: [x(2), x(1) + 1]
```

The two option sets – with and without heuristic – present a reasonable choice in the case that two CPUs are available, but there is not much known about the ideal. Usually the computation time for these settings differ very much: the heuristic is designed to select the apparently best settings in a very progressive way. Indeed, it is not always possible to decide without further experiments which option setting would lead to smallest computation time. Hence, any a priori choice will lead to a suboptimal performance (like all attempts in this area). Because there is much experience behind those heuristic strategies, it will still help to achieve good results in many cases. Also, it yields the best out of the box experience (as far as it is possible here). Disabling the heuristics will usually result in an algorithm which is much closer to the original Buchberger algorithm, but without the overhead of the heuristic functionality itself.

Another important alternative option set consists of computing with and without (dense) linear algebra.

```
In [1]: from polybori.parallel\
    import groebner_basis_first_finished

In [2]: groebner_basis_first_finished([x(2)*x(1), x(1)+1],
   ...:     {'faugere':False,
   ...:     'linear_algebra_in_last_block': False},
   ...:     {'faugere':True,
   ...:     'linear_algebra_in_last_block': True})
Out[2]: [x(2), x(1) + 1]
```

This parallelization, i. e. running different strategies in parallel, seems to be promising because it potentially provides a nonlinear speedup (higher factor than number of CPUs). This can also be observed in other domains: for instance, current SAT-solvers make use of this by choosing different random seeds.

On the other hand, this approach does not contain any cooperation between the threads/processes. At best it is just as fast as the best variant. When studying a particular kind of systems, there is usually a good guess in advance for the best strategy. In this setting, it would be better to use all available CPUs in order to apply the guessed variant to suitable subproblems. The used data structure in PolyBoRi and the functional approach for handling them, seems to be suited to a cooperative shared memory parallelization. However, the underlying library CUDD does not provide any support for that. In this way, cooperative parallelization of ZDD operations remains a challenge for the future. While the ZDD operations cannot be parallelized easily, the linear algebra backend using M4RI [Albrecht and Bard, 2008, Albrecht et al., 2008] can already make use of several CPUs (when compiled with appropriate settings).

7.1.6.2 Elimination of variables

Given Boolean generators G of an ideal

$$I := \langle G \cup \{x_1^2 + x_1, \dots, x_n^2 + x_n, y_1^2 + y_1, \dots, y_m^2 + y_m\}\rangle \subset \mathbb{Z}_2[x_1, \dots, x_n, y_1, \dots, y_m]$$

we would like to compute a generating system H of Boolean polynomials where

$$\langle H \cup \{y_1^2 + y_1, \dots, y_m^2 + y_m\}\rangle = I \cap \mathbb{Z}_2[y_1, \dots, y_m].$$

This can be done as in the classical case despite the field equations using an elimination ordering for $x_1, \dots, x_n$.

Example

```
In [1]: declare_ring([Block("x",3),Block("y",3)],globals())
Out[1]: <polybori.dynamic.PyPolyBoRi.Ring object at 0x1848b10>

In [2]: change_ordering(block_dp_asc)

In [3]: for b in block_start_hints:
   ...:     append_ring_block(b)


In [4]: G=[x(0)*x(2)*y(0)*y(1) + y(1)*y(2) + y(1),
   ...:  x(1)*x(2)*y(0)*y(2) + x(0)*x(1)*y(2) + y(1),
   ...:  x(0)*x(1)*x(2)*y(1) + x(1) + y(1)*y(2)]

In [5]: H=groebner_basis(G)

In [6]: H
Out[6]:
[x(0)*y(0)*y(1) + x(0)*y(1) + y(0)*y(1) + y(1),
 x(1) + y(1),
 x(2)*y(1) + x(0)*y(1) + y(1),
 y(1)*y(2) + y(1)]

In [7]: [p for p in H
    if p.set().navigation().value()>=y(0).index()]
Out[7]: [y(1)*y(2) + y(1)]
```

For special cases elimination (depending on the formulation of the equations) elimination of (auxiliary) variables can be done much faster as can be seen in 7.2.5.3.

7.2 How to program efficiently

The goal of this section is to explain how to get most performance out of PolyBoRi using the underlying ZDD structure. This awareness can be seen on several levels

- ZDD unaware, pure algebraic programming
- low-level friendly programming
- replacing algebraic operations by (a composition of) set operations
- decision-diagram style recursive programming with and without caching
- using ZDDs for many other things than polynomial arithmetics

7.2.1 Low level friendly programming

PolyBoRi is implemented as layer over a decision diagram library (currently CUDD).

In CUDD every diagram node is unique: If two diagrams have the same structure, they are in fact identical (same position in memory). Another observation is that CUDD tries to build a functional style API in the C programming language. This means that no existing data is manipulated, only new nodes are created. Functional programming is a priori well suited for caching and multi-threading (at the moment however threading is not possible in PolyBoRi). The ite-operator is the most central function in CUDD. It takes two nodes/diagrams t, e and an index i and creates a diagram with root variable index i and then-branch t, else-branch e. It is necessary that the branches have root variables with bigger index (or are constant). Either, it creates exactly one node, or it retrieves the correct node from the cache. Function calls which come essentially down to a single ite call are very cheap. For example taking the product

$$x_1 \star (x_2 \star (x_3 \star (x_4 \star x_5)))$$

is much cheaper than

$$((((x_1 \star x_2) \star x_3) \star x_4) \star x_5).$$

In the first case, in each step a single node is prepended to the diagram, while in the second case, a completely new diagram is created. The same argument would apply for the addition of these variables. This example shows that having a little bit background about the data structure it is often possible to write code, that looks algebraic and provides good performance as well.

7.2.2 Replace algebra by set operations

Often there exists an alternative description in terms of set operations for algebraic operations which is much faster. This will be illustrated by the following examples.

7.2.2.1 Construct power sets

An example for this behaviour is the calculation of power sets (sets of monomials/polynomials containing each term in the specified variables). The following code constructs such a power set very inefficiently for the first three variables:

```
sum([x(1)**i1*x(2)**i2*x(3)**i3
    for i1 in (0,1)
    for i2 in (0,1)
    for i3 in (0,1)])
```

The algorithm has of course exponential complexity in the number of variables. The resulting ZDD however has only as many nodes as variables (Figure 3.15). In fact it can be constructed directly using the following function (from specialsets.py).

```
def power_set(variables):
    variables=sorted(set(variables),reverse=True,key=key)
    res=Polynomial(1).set()
    for v in variables:
        res=if_then_else(v,res,res)
    return res
```

Note, that we switched from polynomials to Boolean sets. We inverse the order of variable indices for iteration to make the computation compatible with the principles in 7.2.1 (simple ite operators instead of complex arithmetical operations in each step).

7.2.2.2 All monomials of degree d

The following function constructs the complete homogeneous polynomial/BooleSet containing all possible monomials of degree d (Figure 3.17).

```
def all_monomials_of_degree_d(d,variables):
    if d==0:
        return Polynomial(1).set()
    if len(variables)==0:
        return BooleSet()
    variables=sorted(set(variables),reverse=True,key=top_index)
    m=variables[-1]
    for v in variables[:-1]:
        m=v+m
    m=m.set()
    i=0
    res=Polynomial(1).set()
    while(i<d):
        i=i+1
        res=res.cartesian_product(m).diff(res)
    return res
```

We use the set of all monomials of one degree lower using the cartesian product with the set of variables and remove every term where the degree does not increase (Boolean multiplication: $x^2 = x$).

7.2.3 Direct constructions of diagrams

Sometimes, it is possible to construct the decision diagram directly as its structure is known from theoretical results. In the following, we construct the polynomial from section 7.2.2.2 directly by elementary `if_then_else` operations. This will save much recursion overhead.

```
def all_monomials_of_degree_d(d, variables):
    variables=Monomial(variables)
    variables=list(variables.variables())
    if d>len(variables):
        return Polynomial(0)
    if d<0:
        return Polynomial(1)
    if len(variables)==0:
        assert d==0
        return 1
    deg_variables=variables[-d:]
    #this ensures sorting by indices
    res=Monomial(deg_variables)
    ring=Polynomial(variables[0]).ring()
    for i in xrange(1, len(variables)-d+1):
        deg_variables=variables[-d-i:-i]
        res=Polynomial(res)
        nav=res.navigation()
        navs=[]
        while not nav.constant():
            navs.append(BooleSet(nav,ring))
            nav=nav.then_branch()
        acc=Polynomial(1)
        for (nav, v) in reversed(zip(navs, deg_variables)):
            acc=if_then_else(v, acc, nav)
        res=acc
    return res.set()
```

7.2.4 Case study: Graded part of a polynomial

In the following we will show five variants for implementing a function that computes the sum of all terms of degree d in a polynomial f.

7.2.4.1 Simple, algebraic solution

```
def simple_graded(f, d):
    return sum((t for t in f.terms() if t.deg()==d))
```

This solution is obvious, but quite slow.

7.2.4.2 Low level friendly, algebraic solution

```python
def friendly_graded(f, d):
    vec=BoolePolynomialVector()
    for t in f.terms:
        if t.deg()!=d:
            continue
        else:
            vec.append(t)
    return add_up_polynomials(vec)
```

We leave it to the heuristic of the `add_up_polynomials` function how to add up the monomials. For example a divide and conquer strategy is quite good here.

7.2.4.3 Highlevel with set operations

```python
def highlevel_graded(f,d):
    return Polynomial(
        f.set().intersect(
            all_monomials_of_degree_d(
                d,f.vars_as_monomial())))
```

This solution builds on the fast intersection algorithm and decomposes the task in just two set operations which is very good.

However it can be quite inefficient when f has many variables. This can increase the number of steps in the intersection algorithm (which takes with high probability the else branch of the second argument in each step).

7.2.4.4 Recursive

The repeated unnecessary iteration over all variables in f (during the `intersection` call in the last section) can be avoided by taking just integers as second argument for the recursive algorithm (in the last section this was `intersection`).

```python
def recursive_graded(f,d):
    def do_recursive_graded(f,d):
        if f.empty():
            return f
        if d==0:
            if Monomial() in f:
                return Polynomial(1).set()
            else:
                return BooleSet()
        else:
            nav=f.navigation()
            if nav.constant():
                return BooleSet()
            return if_then_else(
                nav.value(),
                do_recursive_graded(
                    BooleSet(
                        nav.then_branch(),
                        f.ring()),d-1),
```

```
                do_recursive_graded(
                    BooleSet(
                        nav.else_branch(),
                        f.ring()),d))
    return Polynomial(
        do_recursive_graded(f.set(),d))
```

Recursive implementations are very compatible with our data structures, so are quite fast. However this implementation does not use any caching techniques. CUDD recursive caching requires functions to have one, two or three parameters which are of ZDD structure (so no integers). We can encode the degree d by the d-th Variable in the Polynomial ring.

7.2.4.5 Decision-diagram style recursive implementation in PolyBoRi

The C++ implementation of the functionality in PolyBoRi is given in this section which is recursive and uses caching techniques.

```
// determine the part of a polynomials of a given degree
template <class CacheType, class NaviType,
class DegType, class SetType>
SetType
dd_graded_part(const CacheType& cache, NaviType navi, DegType deg,
               SetType init) {


  if (deg == 0) {
    while(!navi.isConstant())
      navi.incrementElse();
    return SetType(navi);
  }

  if(navi.isConstant())
    return SetType();

  // Look whether result was cached before
  NaviType cached = cache.find(navi, deg);

  if (cached.isValid())
    return SetType(cached);

  SetType result =
    SetType(*navi,
            dd_graded_part(cache,
                navi.then_branch(), deg - 1, init),
            dd_graded_part(cache,
                navi.else_branch(), deg, init)
            );
```

```
  // store result for later reuse
  cache.insert(navi, deg, result.navigation());

  return result;
}
```

The encoding of integers for the degree as variable is done implicitly by our cache lookup functions.

7.2.5 Case study: Evaluation of a polynomial

7.2.5.1 Substitute a single variable x in a polynomial by a constant c

Given a Boolean polynomial f, a variable x and a constant c, we want to plug in the constant c for the variable x.

Naïve approach

The following code shows how to tackle the problem by manipulating individual terms. While this is a very direct approach, it is quite slow. The method `reducible_by` gives a test for divisibility.

```
def subst(f,x,c):
    if c==1:
        return sum([t/x for t in f.terms()
            if t.reducible_by(x)])+\
            sum([t for t in f.terms()
                if not t.reducible_by(x)])
    else:
        #c==0
        return sum([t for t in f.terms()
            if not t.reducible_by(x)])
```

Solution 1: Set operations

In fact, the problem can be tackled quite efficiently using set operations.

```
def subst(f,x,c):
   i=x.index()
   c=Polynomial(c)#if c was int is now converted mod 2,
   #so comparison to int(0) makes sense
   s=f.set()
   if c==0:
       #terms with x evaluate to zero
       return Polynomial(s.subset0(i))
   else:
       #c==1
       return Polynomial(s.subset1(i))+Polynomial(s.subset0(i))
```

Solution 2: Linear lexicographical lead rewriting systems

Linear lexicographical rewriting systems and algorithm for them have been introduced formally in section 4.3. Plugging in a constant in a polynomial forms a rewriting problem and can be solved by calculating a normal form against a Gröbner basis. In this case the system is $\{x + c\} \cup \{x_1^2 + x_1, \ldots, x_n^2 + x_n\}$ (we assume that $x = x_i$ for some i). So it indeed matches the case that all Boolean polynomials have pairwise different linear leading terms w. r. t. lexicographical ordering. Hence we can use special functions.

First, we encode the system $\{x + c\}$ into one diagram

```
d=ll_encode([x+c])
```

This is a special format representing a set of such polynomials in one diagram which is used by several procedures in PolyBoRi. Then we may reduce f by this rewriting system

```
ll_red_nf_noredsb(f,d)
```

The procedure `ll_red_nf_noredsb` has been introduced under the name llnf in section 4.3. We try to use a little bit longer more explicit names in PolyBoRi than they are appropriate in mathematical formulas. This can be simplified in our special case in two ways.

1. If our system consists of exactly **one** Boolean polynomial, `ll_encode` does essentially a type conversion only (and much overhead). This type conversion can be done implicitly (at least using the `boost::python`-based interface `ipbori`).

   So you may call

   ```
   ll_red_nf_noredsb(f,x+c)
   ```

   In this case, there is no need for calling `ll_encode`. The second argument is converted implicitly to BooleSet.

2. A second optimization is to call just

   ```
   ll_red_nf_redsb(f,x+c)
   ```

   Note that $\{x + c\}$ is a reduced Boolean Gröbner basis: Equivalently

   $$\{x + c, x_1^2 + x_1, \ldots, x_n^2 + x_n\} \backslash \{x^2 + x\}$$

   is a reduced Gröbner basis).

7.2.5.2 Evaluate a polynomial by plugging in a constant for each variable

The next task will be to rewrite a polynomial $f(x_1, \ldots, x_n)$ by rules $x_i \mapsto c_i$ for each $i \in \{1, \ldots, n\}$ where $c_1, \ldots, c_n$ are constants.

Naive approach

First, we show it in a naive way, similar to the first solution above.

```
def evaluate(f,m):
    res=0
    for term in f.terms():
        product=1
        for variable in term.variables():
            product=m[variable]*product
        res=res+product
    return Polynomial(res)
```

Solution 1: n set operations

The following approach is faster as it does not involve individual terms, but set operations.

```
def evaluate(f,m):
   while not f.constant():
       nav=f.navigation()
       i=nav.value()
       v=Variable(i)
       c=m[v]
       if c==0:
           #terms with x evaluate to zero
           f=Polynomial(nav.then_branch())
       else:
           #c==1
           f=Polynomial(nav.then_branch())+
              Polynomial(nav.else_branch())
       return f
```

For example, the call

```
evaluate(x(1)+x(2),{x(1).index():1,x(2).index():0})
```

results in `1`.

We deal here with navigators which is dangerous because they do not increase the internal reference count on the represented polynomial substructure. So, one has to ensure, that f is still valid as long as we use a navigator on f. But it will show its value on optimized code (e. g. Cython) where it causes less overhead. A second point, why it is desirable to use navigators is, that their `then_branch`- and `else_branch`-methods immediately return (without further calculations) the results of the `subset0` and `subset1`-functions when the latter are called together with the top variable of the diagram f. In this example, this is the crucial point in terms of performance. But, since we already call the polynomial construction on the branches, reference counting of the corresponding subpolynomials is done anyway.

This is quite fast, but suboptimal because only the inner functions (additions) use caching. Furthermore, it contradicts the usual ZDD recursion and generates complex intermediate results.

Solution 2: Linear lexicographical lead rewriting systems

The same problem can also be tackled by the linear-lead routines. In the case, when all variables are substituted by constants, all intermediate results (generated during `ll_red_nf_redsb/ll_red_nf_noredsb`) are constant. In general, we consider the overhead of generating the encoding d as small since it consists of very few, tiny ZDD operations only (and some Python overhead in the quite general `ll_encode`).

```
d=ll_encode([x+cx,y+cy])
ll_red_nf_noredsb(f,d)
```

Since the tails of the polynomials in the rewriting system consist of constants only, this forms also a reduced Gröbner basis. Therefore, you may just call

```
ll_red_nf_redsb(f,d)
```

This is assumed to be the fastest way.

7.2.5.3 General linear lexicographical lead rewriting Systems

We used `ll_red_nf_redsb/ll_red_nf_noredsb` functions on rewriting systems where the tails of the polynomials were constant and the leading term linear. They can be used in a more general setting (which allows to eliminate auxiliary variable) where we only demand that the lexicographical leading monomials are pairwise distinct and linear. This was introduced in section 4.3 as *linear lexicographical rewriting systems.*

We know that such systems form together with the complete set of field equations Gröbner bases w. r. t. lexicographical ordering.

In particular we can use `ll_red_nf_redsb` to speedup substitution of a variable x by a value v also in the more general case that the lexicographical leading term of $x + v$ is equal to x. This can be tested most efficiently by the expression

```
x.set().navigation().value()>v.set().navigation().value().
```

In many cases, we have a bigger equation system where many variables have a linear leading term w. r. t. lexicographical ordering (at least one can optimize the formulation of the equations to fulfill these condition). These systems can be handled by the function `eliminate` in the module `polybori.ll`. I returns three results:

1. A maximal subset L of the equation system which forms a linear lexicographical lexicographical rewriting system.

2. A normal form algorithm f, s. th. $f(p)$ forms a reduced normal form of p against the Gröbner basis consisting of L and the field equations.

3. A list of polynomials R which are in reduced normal form against L, s. th. $L \cup R$ spans modulo field equations the same ideal as the original equation system.

```
In [1]: from polybori.ll import eliminate

In [2]: E=[x(1)+1,x(1)+x(2),x(2)+x(3)*x(4)]

In [3]: (L,f,R)=eliminate(E)
```

```
In [4]: L
Out[4]: [x(1) + 1, x(2) + x(3)*x(4)]

In [5]: R
Out[5]: [x(3)*x(4) + 1]

In [6]: f(x(1)+x(2))
Out[6]: x(3)*x(4) + 1
```

7.3 Other techniques

In this section we treat several smaller aspects. We treat the PolyBoRi data format, Boolean interpolation and finally we will show how to reuse the PolyBoRi components to build a small custom Gröbner basis algorithm.

7.3.1 Storing polynomial data in a file

In PolyBoRi we have default file format and tools which read the files and generate references for our test suite.

The format is a normal Python file with a few exceptions:

- It contains a ring declaration, followed by a list of polynomials called `ideal`.
- Most PolyBoRi standard imports are done by default.
- The second parameter for the ring declaration (the `globals()` dictionary) can be omitted.
- The polynomial system `ideal` is supposed to be the main data. Usually we calculate a Gröbner basis of it.

```
declare_ring([Block("x",4, reverse=False)])
ideal=[
x(1)+x(3),
x(0)+x(1)*x(2)]
```

The data file can be loaded using the following commands.

```
In [1]: from polybori.gbrefs import load_file

In [2]: data=load_file("data-sample.py")

In [3]: data.ideal
Out[3]: [x(1) + x(3), x(0) + x(1)*x(2)]
```

7.3.2 Reinterpretation of Boolean sets as subsets of the vector space $\mathbb{Z}_2^n$

Let S be a Boolean set. For example, consider

```
In [1]:S=BooleSet([x(0),x(1)*x(2)])

In [2]:S
Out[2]:{{x(1),x(2)}, {x(0)}}
```

S is a set of sets of variables. Our usual interpretation is to identify it with a polynomial with corresponding terms:

```
In [3]:Polynomial(S)
Out[3]:x(1)*x(2) + x(0)
```

Another interpretation is to map a set of variables m to a vector v in $\mathbb{Z}_2^n$. The i-th entry of v is set to 1 if and only if the i-th variable occurs in m. So we can identify $\{x_0\}$ with $(1, 0, 0)$ and $\{x_1, x_2\}$ with $(0, 1, 1)$ in $\mathbb{Z}_2^3$. Extending this identification from sets of variables to sets of set of variables we can identify S with $\{(1, 0, 0), (0, 1, 1)\}$. Note, that the choice of n as 3 was not determined by S. In fact every bigger n would have also been a candidate. For this reason, some procedures interpreting Boolean sets as subsets of $\mathbb{Z}_2^n$ taking the monomial ambient space as an additional parameter. The full vector space can be constructed by multiplying all needed variables and the set of divisors of the product.

```
In [4]:(x(1)*x(2)*x(3)).divisors()
Out[4]:{{x(1),x(2),x(3)}, {x(1),x(2)},
        {x(1),x(3)}, {x(1)}, {x(2),x(3)}, {x(2)}, {x(3)}, {}}
```

We distinguish between procedures which use subsets of the ambient space (like finding zeros of a polynomial) and procedures which are independent of the ambient space. The first kind of procedures usually gets the ambient space itself, the second kind does not need such an argument.

This encoding has been defined formally in definition 3.1.20.

7.3.2.1 Examples

```
In [1]:f=x(0)+x(1)+x(2)

In [2]:ambient_space=(x(0)*x(1)*x(2)).divisors()

In [3]:ambient_space
Out[3]:{{x(0),x(1),x(2)}, {x(0),x(1)},
        {x(0),x(2)}, {x(0)}, {x(1),x(2)}, {x(1)}, {x(2)}, {}}

In [4]:f.zeros_in(ambient_space)
Out[4]:{{x(0),x(1)}, {x(0),x(2)}, {x(1),x(2)}, {}}

In [5]:S=BooleSet([x(0),x(1)*x(2)])
```

```
In [6]:f.zeros_in(S)
Out[6]:{{x(1),x(2)}}
```

The function `lex_groebner_basis_points` forms an example for the second kind of procedures, those which are independent of the ambient space.

```
In [1]:S=BooleSet([x(0),x(1)*x(2)])

In [2]:from polybori.interpolate import *

In [3]:lex_groebner_basis_points(S,x(0)*x(1)*x(2))
Out[3]:[x(0) + x(2) + 1, x(1) + x(2)]

In [4]:lex_groebner_basis_points(S,x(0)*x(1)*x(2)*x(3))
Out[4]:[x(0) + x(2) + 1, x(1) + x(2), x(3)]
```

This function calculates the reduced lexicographical Gröbner basis of the vanishing ideal of S. Here the ambient space matters as an additional component would mean that the corresponding entries are zero, so we would get an additional generator for the ideal x_3.

7.3.3 Lexicographical normal form of a polynomial against a variety

Let V be a set of points in $\mathbb{Z}_2^n$, f a Boolean polynomial. V can be encoded as a BooleSet as described in 7.3.2. Then we are interested in the normal form of f against the vanishing ideal of V: $\mathrm{I}(V)$. It turns out that the computation of the normal form can be done by the computation of a minimal interpolation polynomial which takes the same values as f on V. The corresponding algorithms have been introduced in section 5.1.

```
In [1]:from polybori.interpolate import *

In [2]:V=(x(0)+x(1)+x(2)+x(3)+1).set()
```

We take $V = \{e_0, e_1, e_2, e_3, 0\}$ where e_i describes the i-th unit vector. For our considerations it does not play any role, if we suppose V to be embedded in $\mathbb{Z}_2^4$ or a vector space of higher dimension.

```
In [3]:V
Out[3]:{{x(0)}, {x(1)}, {x(2)}, {x(3)}, {}}

In [4]:f=x(0)*x(1)+x(1)+x(2)+1

In [5]:nf_lex_points(f,V)
Out[5]:x(1) + x(2) + 1
```

In this case, the normal form of f w. r. t. the vanishing ideal of V consists of all terms of f with degree less or equal to 1.

It can be seen easily that this polynomial forms the same function on V as f. In fact, our computation is equivalent to the direct call of the interpolation function interpolate_smallest_lex which has two arguments: the set of interpolation points mapped to zero and the set of interpolation points mapped to one.

```
In [6]:z=f.zeros_in(V)

In [7]:z
Out[7]:{{x(1)}, {x(2)}}

In [8]:o=V.diff(z)

In [9]:o
Out[9]:{{x(0)}, {x(3)}, {}}

In [11]:interpolate_smallest_lex(z,o)
Out[11]:x(1) + x(2) + 1
```

7.3.4 Partial Boolean functions

A partial Boolean function f is given by two disjoint set of points O and Z. f is defined to have value 1 on O, 0 on Z and is undefined elsewhere. Using the notions of section 5.1 we may write:

$$f = b_{Z_1}^{O_1}$$

For encoding sets of Boolean vectors we use the encoding defined in 7.3.2.

If we identify 1 with the Boolean value True and 0 with False, operations from propositional logic get a meaning for Boolean functions.

We can apply operations like XOR, logical and, and logical or to partial Boolean functions, defined everywhere where the result is uniquely determined on extensions of these functions.

```
In [1]: from polybori.partial import PartialFunction

In [2]: O=BooleSet([Monomial(),x(0)*x(1)])

In [3]: Z=BooleSet([x(2),x(0)*x(2)])

In [4]: f=PartialFunction(zeros=Z,ones=O)

In [5]: f
Out[5]: PartialFunction(
    zeros={{x(0),x(2)}, {x(2)}},
    ones={{x(0),x(1)}, {}})
In [6]: O2=BooleSet([x(1),x(2)])

In [7]: Z2=BooleSet([x(0)*x(1),x(1)*x(2),x(0)*x(2)])

In [8]: g=PartialFunction(zeros=Z2,ones=O2)
```

```
In [9]: f & g
Out[9]: PartialFunction(
          zeros={{x(0),x(1)}, {x(0),x(2)}, {x(1),x(2)}, {x(2)}},
          ones={})

In [10]: f|g
Out[10]: PartialFunction(
           zeros={{x(0),x(2)}},
           ones={{x(0),x(1)}, {x(1)}, {x(2)}, {}})

In [11]: f^g
Out[11]: PartialFunction(zeros={{x(0),x(2)}},
    ones={{x(0),x(1)}, {x(2)}})
```

Since addition of in $\mathbb{Z}_2$ is equivalent to XOR (using this identification with Boolean logic), the operators `&` and `+` coincide.

```
In [12]: f+g
Out[12]: PartialFunction(zeros={{x(0),x(2)}},
    ones={{x(0),x(1)}, {x(2)}})
```

We have also build our interpolation functions as method for our `PartialFunction` class which is a more convenient way to use it.

```
In [13]: f.interpolate_smallest_lex()
Out[13]: x(2) + 1

In [14]: g.interpolate_smallest_lex()
Out[14]: x(0) + x(1) + x(2)
```

7.3.5 Building your own Gröbner basis algorithm

The central class for writing your own Gröbner bases algorithm is called `GroebnerStrategy`. It represents a system of generators (Boolean polynomials) and contains information about critical pairs as well as some extra information like the set of leading terms belonging to these generators.

The most important operations are:

- adding a generator,
- iterating over the system and accessing generators via indices,
- accessing generators via their leading term (heavily used inside our internal routines),
- calculation of normal form via the method `nf`.

After construction several options can be set, e. g. `opt_red_tail` for tail reductions (affects also the `nf` method). The `GroebnerStrategy` keeps track not only of the single generators, but also of properties of the whole system:

- critical pairs (managed automatically),
- leading terms,
- minimal leading terms.

7.3.5.1 Adding a Generator

There are several methods for adding a generator to a `GroebnerStrategy`. It may not contain two generators with the same leading monomial. In this way generators can be accessed with both their index and their leading term.

```
In [1]: g=GroebnerStrategy()

In [2]: g.add_generator(x(1))

In [3]: g[x(1)]
Out[3]: x(1)

In [4]: g.add_generator(x(1)+1)
-----------------------------------------------
ValueError     Traceback (most recent call last)

/Users/michael/sing/PolyBoRi/<ipython console> in <module>()

ValueError: strategy contains already
    a polynomial with same lead
```

An alternative is to push the generator to the (generalized) set of critical pairs instead of adding it directly

```
In [5]: g.add_generator_delayed(x(1)+1)
```

Due to the absence of other pairs, in this example the polynomial is on top of the pair queue.

```
In [6]: g.next_spoly()
Out[6]: x(1) + 1
```

An alternative approach is to let PolyBoRi decide whether a generator is to be added directly to the system or not.

```
In [1]: g=GroebnerStrategy()

In [2]: g.add_as_you_wish(x(1))
```

7.3.5.2 Interreduction

The `interred`-function gives back a system generating the same ideal where no two leading terms coincide. Also, using the parameter `completely` ensures that no leading term divides the terms in the tails of the other generators. Even more, it is easier than the Buchberger algorithm because no critical pairs have to be handled. Actually the `GroebnerStrategy` applies some criteria when adding a generator which produces some overhead. The algorithm works by passing the sorted generators to the strategy. If a generator is (lead) rewriteable, it is rewriteable by generators with smaller leading terms. So it will be rewritten by this procedure. The algorithm stops when no generator is lead rewriteable any more (`completely=False`) or rewriteable (`completely=True`).

```
def interred(l,completely=False):
    """computes a new generating system (g1, ...,gn),
    spanning the same ideal modulo field equations.
    The system is interreduced: For i!=j:
    gi.lead() does not divide any term of gj
    """
    l=[Polynomial(p) for p in l if not p==0]
    l_old=None
    l=tuple(l)
    while l_old!=l:
        l_old=l
        l=sorted(l,key=Polynomial.lead)
        g=ReductionStrategy()
        if completely:
            g.opt_red_tail=True
        for p in l:
            p=g.nf(p)
            if not p.is_zero():
                g.add_generator(p)
        l=tuple([e.p for e in g])
    return list(l)
```

7.3.5.3 A minimal Buchberger algorithm

In this section the `buchberger` function from the module `simplebb` is presented. Unlike PolyBoRi's more sophisticated routines this procedure was developed for educational purposes only:

```
def buchberger(l):
    "calculates a (non minimal) Groebner basis"
    l=interred(l)
    #for making sure, that every polynomial has a
        different leading term
    #needed for add_generator
    g=GroebnerStrategy()
    for p in l:
        g.add_generator(p)
    while g.npairs()>0:
        g.clean_top_by_chain_criterion()
        p=g.next_spoly()
```

```
        p=g.nf(p)
        if not p.is_zero():
            g.add_generator(p)
    return list(g)
```

The criteria are handled by the add_generator-method for immediately applicable criteria and by the function clean_top_by_chain_criterion for the chain criterion.

7.3.5.4 Estimating the number of solutions

This section presents how to use the blocks for building Buchberger algorithms for other tasks like estimating the number of solutions.

First, we observe the following:

- By implicit use of the field equations the number of solutions is always finite.
- We consider finitely many monomials (as we have a degree bound of one per variable).
- The number of standard monomials (monomials not present in the leading ideal) is equal to the number of solutions.
- The leading ideal (w. r. t. to the increasing set of generators) growths monotonously in each step.

This gives a break condition for the number Buchberger algorithm. It is clear at a certain point of the computations that no more than n solutions exist. However, if there are more than n solutions, the full Gröbner basis is computed by the algorithm presented here.

```
def less_than_n_solutions(ideal,n):
    l=interred(ideal)
    g=GroebnerStrategy()
    all_monomials=Monomial([Variable(i) for i
        in xrange(number_of_variables())]).divisors()
    monomials_not_in_leading_ideal=all_monomials
    for p in l:
        g.add_generator(p)
    while g.npairs()>0:
        monomials_not_in_leading_ideal =\
            monomials_not_in_leading_ideal\
            % g.reduction_strategy.minimal_leading_terms
        if len(monomials_not_in_leading_ideal)<n:
            return True
        g.clean_top_by_chain_criterion()
        p=g.next_spoly()
        p=g.nf(p)
        if not p.is_zero():
            g.add_generator(p)
```

```
monomials_not_in_leading_ideal =\
    monomials_not_in_leading_ideal\
    % g.reduction_strategy.minimal_leading_terms
if len(monomials_not_in_leading_ideal)<n:
    return True
else:
    return False
```

Bibliography

H. Abelson, G. J. Sussman, and J. Sussman. *Structure and Interpretation of Computer Programs.* MIT Press, 2 edition, 1996.

D. Abrahams and the Boost Team. *The Boost Library – Version 1.38.* The Boost Team, 2009. URL `http://www.boost.org`.

S. Agnarsson, A. Kandri-Rody, D. Kapur, P. Narendran, and B. Saunders. The complexity of testing whether a polynomial ideal is nontrivial. In *Third MACSYMA User's Conference*, pages 452–458, Schenectady, NY, 1984.

M. Albrecht. Algebraic Attacks on the Courtois Toy Cipher. Diplomarbeit, Universität Bremen, 2006.

M. Albrecht. anf2cnf, 2009. URL `http://bitbucket.org/malb/algebraic_attacks/src/17bd689ccfbf/anf2cnf.py`.

M. Albrecht and G. Bard. *The M4RI Library – Version 20080901.* The M4RI Team, 2008. URL `http://m4ri.sagemath.org`.

M. Albrecht and C. Cid. Algebraic techniques in differential cryptanalysis. In O. Dunkelman, editor, *Proceedings of Fast Software Encryption (FSE) 2009*, volume 5665 of *Lecture Notes in Computer Science*, pages 193–208. Springer, 2009.

M. Albrecht, G. Bard, and W. Hart. Efficient multiplication of dense matrices over GF(2). *CoRR*, abs/0811.1714, 2008. URL `http://arxiv.org/abs/0811.1714`.

J. Allen and S. Collison. Ohloh – free public directory of open source software and people, 2004. URL `http://www.ohloh.net/`.

D. Andres. Algorithms for the computation of b-functions in algebraic D-module theory. Diplomarbeit, RWTH Aachen, in preparation.

F. Armknecht, C. Carlet, P. Gaborit, S. K. W. Meier, and O. Ruatta. *Advances in Cryptology - EUROCRYPT 2006*, volume 4004/2006 of *Lecture Notes in Computer Science*, chapter Efficient Computation of Algebraic Immunity for Algebraic and Fast Algebraic Attacks, pages 147–164. Springer, Berlin/Heidelberg, Germany, 2006.

G. S. Avrunin. Symbolic model checking using algebraic geometry. In Rajeev Alur and Thomas A. Henzinger, editors, *Proceedings of the Eighth International Conference on Computer Aided Verification CAV*, volume 1102, pages 26–37, New

Brunswick, NJ, USA, / 1996. Springer Verlag. URL citeseer.ist.psu.edu/avrunin96symbolic.html.

O. Bachmann and H. Schönemann. Monomial Representations for Gröbner Bases Computations. In *Proc. of the International Symposium on Symbolic and Algebraic Computation (ISSAC'98)*, pages 309–316. ACM Press, 1998.

M. Bardet, J.-C. Faugère, and B. Salvy. Complexity of Gröbner basis computation for semi-regular overdetermined sequences over F_2 with solutions in F_2. In P. Gianni, editor, *Mega 2005 Sardinia (Italy)*, pages 71–75, 2005.

M. Bayer. Sqlalchemy, 2009. URL http://www.sqlalchemy.org/.

B. Becker, R. Drechsler, and R. Werchner. On the relation between BDDs and FDDs. *Inf. Comput.*, 123(2):185–197, 1995. ISSN 0890-5401. doi: 10.1006/inco.1995.1167. URL http://dx.doi.org/10.1006/inco.1995.1167.

T. Becker and V. Weispfennig. *Gröbner bases, a computational Approach to Commutative Algebra.* Graduate Texts in Mathematics, Springer Verlag, 1993.

S. Behnel, R. Bradshaw, and D. S. Seljebotn. Cython – C-extensions for Python, 2009. URL http://www.cython.org.

D. Benz, F. Eisterlehner, A. Hotho, R. Jäschke, B. Krause, and G. Stumme. Managing publications and bookmarks with bibsonomy. In C. Cattuto, G. Ruffo, and F. Menczer, editors, *HT '09: Proceedings of the 20th ACM Conference on Hypertext and Hypermedia*, pages 323–324, New York, NY, USA, June 2009. ACM. ISBN 978-1-60558-486-7. doi: 10.1145/1557914.1557969. URL http://portal.acm.org/citation.cfm?doid=1557914.1557969#.

B. Bérard, M. Bidoit, F. Laroussine, A. Petit, L. Petrucci, P. Schoenebelen, and P. McKenzie. *Systems and software verification: model-checking techniques and tools.* Springer-Verlag New York, Inc., New York, NY, USA, 1999. ISBN 3-540-41523-8.

A. Biere. AIGER, 2007. URL http://fmv.jku.at/aiger/. AIGER is a format, library and set of utilities for And-Inverter Graphs (AIGs).

A. Biere, M. J. H. Heule, H. van Maaren, and T. Walsh, editors. *Handbook of Satisfiability*, volume 185 of *Frontiers in Artificial Intelligence and Applications.* IOS Press, February 2009. ISBN 978-1-58603-929-5.

W. Bosma, J. Cannon, and C. Playoust. The Magma algebra system I. *Journal of Symbolic Computation, 24, 3/4*, pages 235–265, 1997.

M. Brickenstein. Neue Varianten zur Berechnung von Gröbnerbasen. Diplomarbeit, Fachbereich Mathematik, Technische Universität Kaiserslautern, 2004.

M. Brickenstein. Slimgb: Gröbner bases with slim polynomials. *Revista Matemática Complutense*, 23, issue 2:453–466, 2010. doi: 10.1007/s13163-009-0020-0. URL http://www.springerlink.com/content/e4616125w56k5824. The final publication is available at www.springerlink.com.

M. Brickenstein and A. Dreyer. Gröbner-free normal forms for Boolean polynomials. In *ISSAC '08: Proceedings of the twenty-first International Symposium on Symbolic and Algebraic Computation*, pages 55–62, New York, NY, USA, 2008. ACM. ISBN 978-1-59593-904-3. doi: http://doi.acm.org/10.1145/1390768.1390779.

M. Brickenstein and A. Dreyer. PolyBoRi: A framework for Gröbner-basis computations with Boolean polynomials. *Journal of Symbolic Computation*, 44(9):1326–1345, 2009a. ISSN 0747-7171. doi: 10.1016/j.jsc.2008.02.017. URL `http://dx.doi.org/10.1016/j.jsc.2008.02.017`. Effective Methods in Algebraic Geometry.

M. Brickenstein and A. Dreyer. PolyBoRi mercurial repository, 2009b. URL `http://bitbucket.org/brickenstein/polybori/`.

M. Brickenstein, S. Bulygin, S. King, V. Levandovskyy, and G. M. Diaz Toca. Examples for slimgb, 2006. URL `http://www.mfo.de/organisation/institute/brickenstein/groebner/examples.html`.

M. Brickenstein, A. Dreyer, G.-M. Greuel, M. Wedler, and O. Wienand. New developments in the theory of Gröbner bases and applications to formal verification. *Journal of Pure and Applied Algebra*, 213(8):1612–1635, Aug. 2009. ISSN 0022-4049. doi: 10.1016/j.jpaa.2008.11.043. URL `http://dx.doi.org/10.1016/j.jpaa.2008.11.043`. Theoretical Effectivity and Practical Effectivity of Gröbner Bases.

R. E. Bryant. Graph-based algorithms for Boolean function manipulation. *IEEE Transactions on Computers*, 35(8):677–691, 1986.

B. Buchberger. A Criterion for Detecting Unnecessary Reductions in the Construction of a Gröbner Basis. In N. K. Bose, editor, *Recent trends in multidimensional system theory*. 1985.

B. Buchberger. Ein Algorithmus zum Auffinden der Basiselemente des Restklassenrings nach einem nulldimensionalen Polynomideal. Dissertation, Universität Innsbruck, 1965.

S. Bulygin and M. Brickenstein. Obtaining and solving systems of equations in key variables only for the small variants of AES. *Mathematics in Computer Science*, 2010. doi: 10.1007/s11786-009-0020-y. URL `http://www.springerlink.com/content/k486617281521541`. The final publication is available at www.springerlink.com.

M. Caboara, M. Kreuzer, and L. Robbiano. Efficiently Computing Minimal Sets of Critical Pairs. *Journal of Symbolic Computation, 38*, pages 1169–1190, 2004.

R. Cameron. citeulike – free service for managing and discovering scholarly references, 2004. URL `http://www.citeulike.org/`.

F. Chai, X.-S. Gao, and C. Yuan. A characteristic set method for solving boolean equations and applications in cryptanalysis of stream ciphers. *MM-Preprints*, 26, 2008. URL `http://www.mmrc.iss.ac.cn/pub/mm26.pdf/4-csz2-mm.pdf`.

C. Cid, S. Murphy, and M. J. B. Robshaw. Small scale variants of the AES. In H. Gilbert and H. Handschuh, editors, *Fast Software Encryption 2005*, volume 3557 of *Lecture Notes in Computer Science*, pages 145–162. Springer-Verlag, 2005.

J. Clark and D. A. Holton. *A First Look at Graph Theory.* World Scientific, 1991.

M. Clegg, J. Edmonds, and R. Impagliazzo. Using the groebner basis algorithm to find proofs of unsatisfiability. *Proceedings of the Twenty-eighth Annual ACM Symposium on the Theory of Computing*, pages 174–183, 1996.

C. Condrat and P. Kalla. A Gröbner basis approach to CNF-formulae preprocessing. In *Tools and Algorithms for the Construction and Analysis of Systems*, volume 4424 of *Lecture Notes in Computer Science*, pages 618–631. Springer, 2007. doi: 10.1007/978-3-540-71209-1_48. URL `http://www.springerlink.com/content/m406t6m40357x847`.

N. Courtois. How fast can be algebraic attacks on block ciphers? *Cryptology ePrint Archive, Report 2006/168*, 2006. URL `http://eprint.iacr.org/2006/168.pdf`.

N. Courtois and G. Bard. *Algebraic Cryptanalysis of the Data Encryption Standard.*, volume 4887 of *Lecture Notes in Computer Science*, pages 152–169. Springer, 2008. ISBN 3-540-77271-5.

D. Cox, J. Little, and D. O'Shea. *Ideals, Varieties, and Algorithms.* Springer, 3rd edition, 2007.

L. Dagum and R. Menon. OpenMP: an industry standard API for shared-memory programming. *IEEE Computational Science and Engineering*, 5(1):46–55, 1998. URL `http://ieeexplore.ieee.org/xpls/abs_all.jsp?arnumber=660313`.

M. Davis and H. Putnam. A computing procedure for quantification theory. *J. ACM*, 7 (3):201–215, 1960. ISSN 0004-5411. doi: http://doi.acm.org/10.1145/321033.321034.

N. Eén and N. Sörensson. An extensible SAT-solver. In E. Giunchiglia and A. Tacchella, editors, *SAT*, volume 2919 of *Lecture Notes in Computer Science*, pages 502–518. Springer, 2003.

J.-C. Faugère. A new Efficient Algorithm for Computing Gröbner Bases (F_4). *Journal of Pure and Applied Algebra*, 139(1–3):61–88, 1999.

J.-C. Faugère. A new Efficient Algorithm for Computing Gröbner Bases Without Reduction to Zero (F_5). In *Proc. of the International Symposium on Symbolic and Algebraic Computation (ISSAC'02)*, pages 75–83. ACM Press, 2002.

J.-C. Faugère. FGb/Maple interface, 2006. URL `http://fgbrs.lip6.fr/salsa/Software/`.

J.-C. Faugère. Algebraic Cryptanalysis of Hidden Field Equation (HFE) Cryptosystems Using Gröbner Bases. In *Advances in Cryptology - CRYPTO 2003, Lecture Notes in Computer Science 2729/2003*, pages 44–60, 2003.

J.-C. Faugère, P. Gianni, D. Lazard, and T. Mora. Efficient computation of zero-dimensional Gröbner bases by change of ordering. *Journal Symbolic Computation*, 16(4):329–344, 1993. ISSN 0747-7171. doi: http://dx.doi.org/10.1006/jsco.1993.1051.

J. Gago-Vargas, I. Hartillo-Hermoso, J. Martín-Morales, and J. M. Ucha-Enríquez. *Sudokus and Gröbner Bases: Not Only a Divertimento*, pages 155–165. Springer, 2006. doi: 10.1007/11870814_13. URL `http://www.springerlink.com/content/w222g687w8107618`.

E. R. Gansner. Drawing graphs with GraphViz. Technical report, AT&T Bell Laboratories, Murray Hill, NJ, USA, Apr. 2003. URL `http://www.graphviz.org/pdf/libguide.pdf`.

E. R. Gansner, E. Koutsofios, and S. North. Drawing graphs with Dot. Technical report, AT&T Bell Laboratories, Murray Hill, NJ, USA, Jan. 2006. URL `http://www.graphviz.org/pdf/dotguide.pdf`.

S. Gao. Counting Zeros over Finite Fields with Gröbner Bases. master thesis, Carnegie Mellon University, 2009.

M. Gardner. Mathematical games: On cellular automata, self-reproduction, the Garden of Eden and the game of 'Life'. *Scientific American*, 224(2):112–117, Feb. 1971. ISSN 0036-8733.

V. P. Gerdt and M. V. Zinin. A Pommaret division algorithm for computing Gröbner bases in Boolean rings. In *ISSAC '08: Proceedings of the twenty-first International Symposium on Symbolic and Algebraic Computation*, pages 95–102, New York, NY, USA, 2008. ACM. ISBN 978-1-59593-904-3. doi: http://doi.acm.org/10.1145/1390768.1390784.

M. Ghasemzadeh. *A new algorithm for the quantified satisfiability problem, based on zero-suppressed binary decision diagrams and memoization.* PhD thesis, University of Potsdam, Potsdam, Germany, Nov. 2005. URL `http://opus.kobv.de/ubp/volltexte/2006/637/`.

A. Giovini, T. Mora, G. Niesi, L. Robbiano, and C. Traverso. One sugar cube, please or Selection strategies in Buchberger algorithms. In S. Watt, editor, *Proceedings of the 1991 International Symposium on Symbolic and Algebraic Computations, ISSAC'91*, pages 49–54. ACM press, 1991.

D. Gligoroski, V. Dimitrova, and S. Markovski. Quasigroups as Boolean Functions, Their Equation Systems and Gröbner Bases. In M. Sala, T. Mora, L. Perret, S. Sakata, and C. Traverso, editors, *Gröbner Bases, Coding, and Cryptography*, pages 415–420. Springer, 2009. doi: 10.1007/978-3-540-93806-4_31. URL `http://www.springerlink.com/content/j7q0u503h5074837`.

C. P. Gomes, A. Sabharwal, and B. Selman. Model counting. In A. Biere, M. J. H. Heule, H. van Maaren, and T. Walsh, editors, *Handbook of Satisfiability*, volume 185 of *Frontiers in Artificial Intelligence and Applications*, pages 633 – 654. IOS Press, February 2009. ISBN 978-1-58603-929-5.

G.-M. Greuel and G. Pfister. *A SINGULAR Introduction to Commutative Algebra.* Springer Verlag, 2002.

G.-M. Greuel, G. Pfister, and H. Schönemann. SINGULAR – a computer algebra system for polynomial computations. In M. Kerber and M. Kohlhase, editors, *Symbolic computation and automated reasoning, The Calculemus-2000 Symposium*, pages 227–233, Natick, MA, USA, 2001. A. K. Peters, Ltd. ISBN 1-56881-145-4.

G.-M. Greuel, G. Pfister, and H. Schönemann. SINGULAR 3-1-0 — A computer algebra system for polynomial computations, 2009. URL `http://www.singular.uni-kl.de`.

H.-G. Gräbe. The Symbolic Data Project, 2000-2006. `http://www.SymbolicData.org`

E. Guerrini, E. Orsini, and I. Simonetti. *Gröbner Bases, Coding, and Cryptography*, chapter Gröbner Bases for the Distance Distribution of Systematic Codes, pages 367–272. Springer, 2009. doi: 10.1007/978-3-540-93806-4_22. URL `http://www.springerlink.com/content/j7q0u503h5074837`.

K. Hammond. Parallel Functional Programming: An Introduction. In *First Intl. Symposium on Parallel Symbolic Computation.* World Scientific, 1994. URL `http://www-fp.dcs.st-and.ac.uk/~kh/papers/pasco94/pasco94.html`.

O. M. Hansen and J.-F. Michon. Boolean Gröbner basis. In J.-F. Michon, P. Valarcher, and J.-B. Yunès, editors, *Proceedings of BFCA'06 Conference, March 13–15, 2006, Rouen, France*, pages 185–201, 2006.

D. R. Hipp, D. Kennedy, S. Harrelson, and C. Werner. Sqlalchemy, 2009. URL `http://www.sqlite.org/`.

P. Hudak. *The Haskell School of Expression – Learning Functional Programming through Multimedia.* Cambridge University Press, New York, 2000.

IEEE Computer Society. *IEEE Standard Hardware Description Language Based on the Verilog Hardware Description Language.* IEEE Computer Society Press, Piscataway, USA, 1996. ISBN 1-55937-727-5.

D. Ilsen, E. Roebbers, and G.-M. Greuel. *Algebraic and Combinatorial Algorithms for Translinear Network Synthesis*, volume 55, pages 3131–3144. IEEE Circuits and Systems Society, 2008.

J. von zur Gathen and J. Gerhard. *Modern Computer Algebra.* Cambridge University Press, 1999.

A. A. Karatsuba and Y. Ofman. Multiplication of multidigit numbers by automata. *Doklady Akad. Nauk SSSR*, 145:293–294, 1962.

U. Kebschull and W. Rosenstiel. Efficient graph-based computation and manipulation of functional decision diagrams. *Design Automation, 1993, with the European Event in ASIC Design. Proceedings. 4th European Conference on*, pages 278–282, Feb. 1993. doi: 10.1109/EDAC.1993.386463.

U. Kebschull, E. Schubert, and W. Rosenstiel. Multilevel logic synthesis based on functional decision diagrams. *Design Automation, 1992. Proceedings., 3rd European Conference on*, pages 43–47, Mar. 1992. doi: 10.1109/EDAC.1992.205890.

G. Kiczales, J. Lamping, A. Mendhekar, C. Maeda, C. Lopes, J. marc Loingtier, and J. Irwin. Aspect-oriented programming. In *ECOOP*. SpringerVerlag, 1997.

S. A. King. Ideal turaev-viro invariants. *Topology and its Applications*, 154(6): 1141–1156, 2007. ISSN 0166-8641. doi: 10.1016/j.topol.2006.11.005. URL `http://www.sciencedirect.com/science/article/B6V1K-4MR7D3D-1/2/21d102644a1e11cb3d748009e7ae8ee0`.

H. Kitano. Systems biology: A brief overview. *Science*, 295:1662–1664, Mar. 2002.

J. Knappen. *Schnell ans Ziel mit LATEX 2e*. Oldenbourg, München [u.a.], 1997. ISBN 3486241990. URL `http://gso.gbv.de/DB=2.1/CMD?ACT=SRCHA&SRT=YOP&IKT=1016&TRM=ppn+23222059X&sourceid=fbw_bibsonomy`.

D. E. Knuth. *The Art of Computer Programming 3. Sorting and Searching*. Addison-Wesley Longman, 1998.

I. Koren. *Computer Arithmetic Algorithms*. A. K. Peters, Ltd., Natick, MA, USA, 2001. ISBN 1568811608.

U. Krautz, M. Wedler, W. Kunz, K. Weber, C. Jacobi, and M. Pflanz. Verifying full-custom multipliers by Boolean equivalence checking and an arithmetic bit level proof. In *ASP-DAC '08: Proceedings of the 2008 Asia and South Pacific Design Automation Conference*, pages 398–403, Los Alamitos, CA, USA, 2008. IEEE Computer Society Press. ISBN 978-1-4244-1922-7.

R. Laubenbacher and B. Stigler. A Computational algebra approach to the reverse engineering of gene regulatory networks. *Journal of Theoretical Biology*, 229:523–537, 2004.

V. Levandovskyy. Non–commutative computer algebra for polynomial algebras: Gröbner bases, applications and implementation. *Doctoral Thesis, Universität Kaiserslautern*, 2005. Available from http://kluedo.ub.uni-kl.de/volltexte/2005/1883/.

V. Levandovskyy and J. M. Morales. Computational D-module theory with singular, comparison with other systems and two new algorithms. In *ISSAC '08: Proceedings of the twenty-first international symposium on Symbolic and algebraic computation*, pages 173–180, New York, NY, USA, 2008. ACM. ISBN 978-1-59593-904-3. doi: 10.1145/1390768.1390794.

H. Li, I. G. Councill, L. Bolelli, D. Zhou, Y. Song, W. chien Lee, A. Sivasubramaniam, and C. L. Giles. Citeseerx - a scalable autonomous scientific digital library. In *CONFERENCE ON SCALABLE INFORMATION SYSTEMS*. ACM, 2006.

H.-M. Lipp. *Grundlagen der Digitaltechnik*. Oldenbourg, 2002. ISBN 3-486-25916-4.

E. Lucas. Théorie des fonctions numériques simplement périodiques. *American Journal of Mathematics*, 1(2):184–196, 1878. ISSN 00029327. URL http://www.jstor.org/stable/2369308.

M. Mackall. Mercurial, 2009. URL http://mercurial.selenic.com.

A. Martelli, A. Ravenscroft, and D. Ascher. *Python Cookbook.* O'Reilly Media, Inc., March 2005. ISBN 0596007973. URL http://code.activestate.com/recipes/langs/python/.

S. Minato. Zero-Suppressed BDDs for Set Manipulation in Combinatorial Problems. pages 272–277, 1993.

S. Minato. Implicit manipulation of polynomials using zero-suppressed BDDs. In *Proc. of IEEE The European Design and Test Conference (ED&TC'95)*, pages 449–454, Mar. 1995.

A. Mishchenko. *EXTRA v. 2.0: Software Library Extending CUDD Package*, 2003. URL http://web.cecs.pdx.edu/~alanmi/research/extra.htm.

B. Mourrain and O. Ruatta. Relations between roots and coefficients, interpolation and application to system solving. *Journal of Symbolic Computation*, 33(5):679–699, 2002. ISSN 0747-7171. doi: 10.1006/jsco.2002.0530.

R. Oudkerk. Python processing, 2009. URL http://pyprocessing.berlios.de/.

D. Patterson and J. Hennessy. *Computer Organization and Design: The Hardware/software Interface.* Morgan Kaufmann, 2009.

F. Perez and B. E. Granger. IPython: A system for interactive scientific computing. *Computing in Science and Engineering*, 9(3):21–29, 2007. ISSN 1521-9615. doi: http://dx.doi.org/10.1109/MCSE.2007.53.

L. Robbiano. Term Orderings on the Polynomial Ring. In *Proceedings of EUROCAL 85, Lecture Notes in Computer Science* **204**, pages 513–517, 1985.

G. V. Rossum and F. L. Drake. *The Python Language Reference Manual.* Network Theory Ltd., Bristol, United Kingdom, November 2006. ISBN 0954161785.

K. Sakai and Y. Sato. Boolean Gröbner bases. Technical report, 1988. URL http://www.icot.or.jp/ARCHIVE/Museum/TRTM/tm0488.htm.

Y. Sato, A. Nagai, and S. Inoue. *Computer Mathematics*, chapter On the Computation of Elimination Ideals of Boolean Polynomial Rings, pages 334–348. Springer, 2008a. doi: 10.1007/978-3-540-87827-8_29. URL http://www.springerlink.com/content/4885k3712w6q3357.

Y. Sato, A. Suzuki, S. Inoue, and K. Nabeshima. Boolean Groebner bases and sudoku. In *ACA 2008*, 2008b.

F. Somenzi. CUDD: CU decision diagram package. University of Colorado at Boulder, 2005. URL http://vlsi.colorado.edu/~fabio/CUDD/. Release 2.4.1.

M. Soos, K. Nohl, and C. Castelluccia. Extending sat solvers to cryptographic problems. In *SAT '09: Proceedings of the 12th International Conference on Theory and Applications of Satisfiability Testing*, pages 244–257, Berlin, Heidelberg, 2009. Springer-Verlag. ISBN 978-3-642-02776-5. doi: http://dx.doi.org/10.1007/978-3-642-02777-2_24.

W. Stein et al. *Sage Mathematics Software (Version 3.3).* The Sage Development Team, 2009. http://www.sagemath.org.

A. A. Stepanov and M. Lee. The Standard Template Library. Technical Report X3J16/94-0095, WG21/N0482, 1994.

Q. Tran and M. Y. Vardi. Groebner bases computation in Boolean rings for symbolic model checking. In *MOAS'07: Proceedings of the 18th conference on Proceedings of the 18th IASTED International Conference*, pages 440–445, Anaheim, CA, USA, 2007. ACTA Press.

M. Wedler. private communication, 2007.

T. Wichmann. Der FGLM-Algorithmus: verallgemeinert und implementiert in SINGULAR. Diplomarbeit, Fachbereich Mathematik, Technische Universität Kaiserslautern, 1997.

T. Yan. The geobucket data structure for polynomials. *J. Symbolic Computation*, 25(3): 285–294, March 1998.

N. Ziviani, E. S. de Moura, G. Navarro, and R. Baeza-Yates. Compression: A key for next-generation text retrieval systems. *Computer*, 33(11):37–44, Nov. 2000. ISSN 0018-9162. URL http://www.computer.org/computer/co/ry037abs.htm;http://dlib.computer.org/co/books/co2000/pdf/ry037.pdf.

Curriculum vitae

General Data

5/4/1979	Born in Hamburg
Family Status	unmarried

School

1985-1989	Grundschule Niederbieber (Neuwied)
1989-1998	Rhein-Wied Gymnasium Neuwied
1998	Abitur
1998-1999	Civilian Service: DRK-Krankenhaus Neuwied

University

1999-2005	Technische Universität Kaiserslautern: Mathematics with minor subject Computer Science
2001	Acceptance for the Studienstiftung des Deutschen Volkes
2001	Vordiplom
2004	Diplom
1/1/2005-7/31/2005	Graduiertenkolleg Mathematik und Praxis

Professional Life

since 2005	Software developer at the Mathematisches Forschungsinstitut Oberwolfach gGmbH

August 2, 2010

Curriculum vitae

Allgemein

4.5.1979	Geburt in Hamburg
Familienstand	ledig

Schule

1985-1989	Grundschule Niederbieber (Neuwied)
1989-1998	Rhein-Wied Gynmnasium Neuwied
1998	Abitur
1998-1999	Zivildienst: DRK-Krankenhaus Neuwied

Universität

1999-2005	Technische Universität Kaiserslautern: Mathematik mit Nebenfach Informatik
2001	Aufnahme in die Studienstiftung des Deutschen Volkes
2001	Vordiplom
2004	Diplom
1.1.2005-31.7.2005	Graduiertenkolleg Mathematik und Praxis

Berufsleben

seit 2005	Softwareentwickler in der Mathematisches Forschungsinstitut Oberwolfach gGmbH

August 2, 2010